# COSMIC
## MENAGERIE
### A VISUAL JOURNEY THROUGH THE UNIVERSE

WRITTEN AND ILLUSTRATED BY

# MARK A. GARLICK

RED
LEMON
PRESS

First published in Great Britain by
Red Lemon Press Limited
Northburgh House
10 Northburgh Street
London EC1V 0AT, UK

ISBN 978-1-78342-004-9

A CIP catalogue record for this book is
available from the British Library.

Printed and bound in China
9 8 7 6 5 4 3 2 1

www.redlemonpress.com

Red Lemon Press Limited is part of
the Bonnier Publishing Group

www.bonnierpublishing.com

**Take your journey through the
Universe beyond the
printed page!**

Access exclusive animations using
the free *Cosmic Menagerie* app. Simply
download the app from iTunes or Google
Play, point your device at the page
whenever you see the smartphone
symbol, and the animation will launch.
The *Cosmic Menagerie* app requires an
internet connection, and can be used on
iPhone, iPad or Android devices. For
direct links to download the app, visit
**www.redlemonpress.com/apps**

This app is compatible with iPhone 3GS, iPhone 4,
iPhone 4S, iPhone 5, iPod touch (4th generation), iPod
touch (5th generation), iPad 2 Wi-Fi, iPad 2 Wi-Fi + 3G,
iPad (3rd generation), iPad Wi-Fi + 4G, iPad (4th
generation), iPad Wi-Fi + Cellular (4th generation),
iPad mini and iPad mini Wi-Fi + Cellular. Requires iOS
4.3 or later. This app is optimized for iPhone 5. Android
users must be using a device with version 2.2 or later.

# COSMIC MENAGERIE

## A VISUAL JOURNEY THROUGH THE UNIVERSE

WRITTEN AND ILLUSTRATED BY

### MARK A. GARLICK

RED
LEMON
PRESS

# CONTENTS

# INTERSTELLAR FOG

# GALAXIES... AND BEYOND

# FOREWORD

This book proves that imagination is an important tool of science. It is bursting with images of objects that no one has ever seen.

The Hubble Space Telescope (HST) orbits above the Earth's atmosphere, freeing it from the limitation of our burbling air, but even the HST cannot show you the surface of a distant star. Our keenest tool for astronomical imaging has a sharpness of view that is strictly limited by the wavelength of light itself and the finite diameter of its mirror. But here we are shown scenes of the hidden Universe, page after glorious page, vividly depicted. Eye-popping images show the blotchy surfaces of teenage stars, the distorted shapes of whirling dervish stars, the streams of matter flowing in binaries, and even the swirling cyclone of doomed matter sliding towards the boundaries of a black hole.

How does Mark A. Garlick know what these stars and events look like? No one has ever seen them. Are they real? The dazzling images in this book are not based on photographs. They are Garlick's creations. Yet, they are based on real evidence. Astronomers use subtle clues to puzzle out the nature of stars. Garlick has taken these physical models out of the pages of learned astronomical journals and made them concrete, specific and vivid. He views the unseen from impossible perspectives to create images that help you to grasp the nature of these amazing objects.

The laws of physics at play here are very simple: mostly gravity pulling in and gas pressure pushing out, plus a little nuclear fusion to light the scene. Simple atoms create an amazing range of possibilities. Atoms make up only five per cent of the mass and energy of the Universe, but they are the most interesting part. You and I are made of atoms – as are the dazzling astronomical specimens of the *Cosmic Menagerie*. We are not apart from the Universe; we are a part of it.

**Robert P. Kirshner**
Clowes Professor of Science, Harvard University

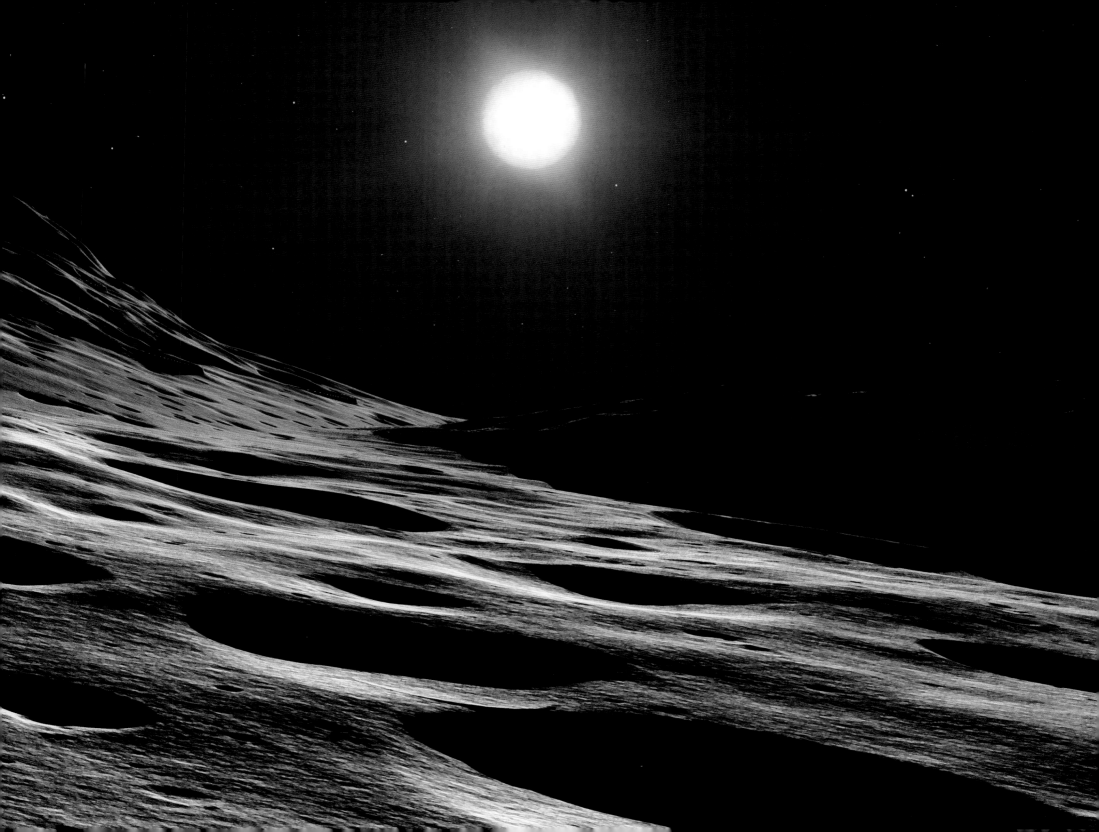

# PLANETS AND RELATED SPECIMENS

*Far above the globe of planet Earth we have an interplanetary visitor. This is a near-Earth asteroid, a chunk of space rock moving around the Sun in an orbit that crosses that of our own planet.*

# INTRODUCTION

Most people have an idea what a planet is. After all, we live on one: our humble Earth. If you glance at the clear night sky, the chances are you will see other planets. Venus and Jupiter are particularly dazzling, but they can be mistaken for bright stars. Up close, though, planets and stars are totally different beasts. Stars (such as the Sun) shine by their own light, but planets emit no light and little heat. We see them merely by the light they reflect from the Sun.

Planets vary enormously from one specimen to another. We are all familiar with the Earth: solid, rocky and generally temperate. But our world is nothing like the largest planets, the gas giants Jupiter and Saturn. Not only are these a good ten times larger, they are entirely fluid, with no surface to set foot on. Then there are worlds such as Mercury: rocky and metallic like the Earth, but one-third of its size.

Related to planets are dwarf planets – a relatively new category of Solar System object. Put simply, astronomers deem them too small to be real planets. One of the dwarf planets, Ceres, is also categorized as another class of Solar System object: an asteroid. Asteroids number in the tens of millions. While a handful measure up to the scale of Ceres, the overwhelming majority are puny, potato-shaped lumps of rock, metal and carbon, orbiting the Sun between Mars and Jupiter.

There is yet more to the Solar System, including comets – great 'dirty snowballs' that orbit the Sun in droves; planetary satellites, such as the Moon; and the mysterious trans-Neptunian bodies that lurk in the outer reaches of the planetary realm. We will meet all these objects in this chapter. First, let's start at the beginning: where did all this stuff come from?

# FORMATION
## BIRTH OF THE SOLAR SYSTEM

The dinosaur age ended 65 million years ago. That time is already far beyond human experience, but the Earth, the Sun and everything in the Solar System is 70 times older still. At 4,550 million years of age, our Solar System is unimaginably ancient. Before then, there was no Sun; no planets. There were only raw materials: a vast nebula of gas and dust called a giant molecular cloud (*see* pp96–97). Molecular clouds are knotty, with some regions denser than others. In these places, gravity pulls inwards and the knots become tighter still. After two million years, so-called globules develop – roughly spherical shells of gas and dust with a dense central core. Nearly five billion years ago, inside a cocoon like this, the Sun and the planets were forged.

Gradually, the solar globule contracted further under gravity, warming as it did so. As it shrank, it spun faster and faster, and as it wound up, it flattened out. After 100,000 more years, the globule was transformed. In its place was a swirling, dusty pancake called the Solar Nebula. In its hot central region was a glowing mass, bigger than the present orbit of Mercury, called a protostar (*see* pp34–35).

Millions of years later, with continued contraction and warming, the protostar formed the Sun. Meanwhile, out in the disc, tiny dust particles stuck together, attracted by electrostatic forces. They grew bigger and bigger, into objects the size of pebbles, then rocks and then boulders. Still the collisions continued, until the boulders grew into mountain-sized planetary building blocks called planetesimals. The impact rate stepped up a gear as these large objects collided no longer by chance, but because of their mutual gravitational attractions. After a few more million years, most planetesimals had vanished, swept into eight major planets and countless asteroids – debris from the formation of the Solar System that survives to this day.

# GAS GIANTS
## JUPITER AND SATURN

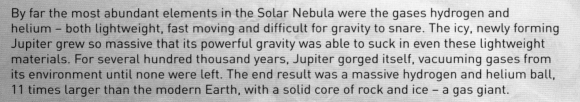

In the Solar Nebula, billions of years ago, there was a strong temperature gradient. Close to the newly forming Sun, where it was hottest, icy substances were unable to condense (turn from a gas into a liquid or solid), while substances such as iron and rock were readily able to do so. Roughly five times further from the Sun than the present orbit of the Earth, at the so-called snow line, the temperature dropped to a point where ices could condense. Icy particles are much stickier than the rocky and metallic ones that condensed closer to the Sun, so the snow-line planetesimals quickly grew to planet-sized dimensions. In just 100,000 years or so, an icy kernel 15 times more massive than the Earth appeared: Jupiter was forming.

By far the most abundant elements in the Solar Nebula were the gases hydrogen and helium – both lightweight, fast moving and difficult for gravity to snare. The icy, newly forming Jupiter grew so massive that its powerful gravity was able to suck in even these lightweight materials. For several hundred thousand years, Jupiter gorged itself, vacuuming gases from its environment until none were left. The end result was a massive hydrogen and helium ball, 11 times larger than the modern Earth, with a solid core of rock and ice – a gas giant.

Twice as far from the Sun, another gas giant formed: Saturn. Being further out, where the density of raw materials was much less, Saturn could not grow to the dimensions of its larger cousin. Still, it is a true giant – a respectable nine times larger than the Earth.

Gas giants are the largest planets we know of. Jupiter and Saturn are the only two in our Solar System, but elsewhere in space, orbiting other stars, these planets are very common indeed.

Jupiter is the quintessential gas
giant planet. It has no solid surface,
just a vast, colourful envelope of
super-compressed hydrogen and
helium surrounding a solid core of
rock and ice about the size of the Earth.

Scan the page to view animation

Neptune's colour is derived from methane in its stormy, turbulent atmosphere, which absorbs light from the red part of the spectrum, leaving blue to be seen. A system of narrow rings encircles the planet.

# ICE GIANTS
## URANUS AND NEPTUNE

The next planets off the production line were Uranus and Neptune, at even greater distances from the Sun than the gas giants. Let's pause to put these great expanses into perspective. Astronomers use astronomical units (AU) to measure distances in the Solar System. One AU is the Earth–Sun distance (150 million kilometres). Today, Jupiter orbits the Sun at an average distance of 5.2 AU. Saturn, by contrast, orbits at 9.2 AU. Uranus is much further out, at 19.2 AU, and Neptune is right on the edge of the realm of the planets, fully 30 times further from the Sun than the Earth.

At these great distances from the Sun, more than four billion years ago, the material in the Solar Nebula was widely spread. Plus, the time taken to orbit the Sun at these enormous distances is much greater than, for example, the orbital period of Saturn. Both these circumstances meant that Uranus and Neptune grew much more slowly than their giant siblings closer to the Sun. While Saturn and Jupiter reached full size by about three million years after the appearance of the Solar Nebula, the growth of Uranus and Neptune took roughly three times longer. They still ended up with only five per cent of the mass of mighty Jupiter.

Still, Uranus and Neptune are sizeable worlds. They are almost equal in size and mass: 17 and 15 times heavier than the Earth respectively and four times its diameter. However, they lack the extensive hydrogen and helium mantles of the gas giants. While Jupiter is 90 per cent hydrogen, this gas makes up only about 20 per cent of Uranus and Neptune. In place of the hydrogen are ices of ammonia and methane. As a result, Uranus and Neptune are regarded as a separate class of planets – the ice giants.

*Uranus is almost exactly the same size as Neptune. It is a relatively featureless planet, lacking the clouds and storms of Neptune.*

# TERRESTRIALS
## PLANETS OF ROCK AND IRON

The final planets to emerge from the chaos of the Solar Nebula were the smaller ones, of which our Earth is the largest. They are known as terrestrial or telluric planets – from the Latin words for Earth: *tellus*, *terrestris* or *terra*. However, the other three terrestrial planets are nothing like the Earth. The term 'terrestrial' is used merely to describe planets with solid surfaces of rocky and metallic materials.

The growth of the terrestrial worlds – Mercury, Venus, Earth and Mars – was slow for two reasons. First, the rocks and metals from which these worlds formed were much less abundant than the gases and ices that formed the giant planets. Second, fragments of colliding rock or metal are more likely to bounce off each other than colliding chunks of ice, because the latter have better adhesive properties. So, although the initial emergence of rocky and metallic planetesimals close to the Sun was fairly rapid, it took much longer for them to grow into fully fledged worlds. After perhaps ten million years, by which time the giant planets were fully grown, four dominant protoplanets populated the inner Solar System, destined to become terrestrial worlds. Even so, it took around ten times longer for these worldlets to mop up the remaining planetesimal debris orbiting the Sun and grow to full size.

The terrestrial planets vary in scale considerably. Earth is the largest and most massive, while Venus is almost exactly the same size. Mars is only half the diameter of the Earth and Mercury is only one-third the size – and made almost entirely of solid iron. Astronomers think that Mercury, the closest planet to the Sun, was originally much larger and may have lost much of its mass during a collision with a similar-sized world billions of years ago. This may explain its heavy metal content – its outer rocky shell might have been blasted away in the impact.

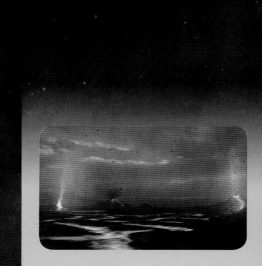

*Earth is shown here in a primordial state, in the process of formation. It is a fiery, turbulent world.*

 **Scan the page to view animation**

Mars is the second smallest of the terrestrial planets. Its surface is peppered with craters, great chasms and gigantic volcanoes – the largest in the known Solar System. The planet's red colour comes from the presence of rust, or iron oxide, which blankets the dusty, Martian soil.

"...constant gravitational bullying ensures that the asteroids will remain, probably forever, individuals..."

This is a binary asteroid – not one, but two, each orbiting the other like a vast dumb-bell in space, invisibly tethered by gravity. Asteroids with satellites are quite common, although usually one component is much larger than the other.

*Scan the page to view animation*

# ASTEROIDS
## SOLAR SYSTEM DEBRIS

Given the fiery birth of the Solar System, it would be surprising if there were no remaining debris from that stormy period. Indeed, nearly five billion years later, the Solar System is stuffed with leftover chunks of rock, metal and other materials. These primordial fragments of the Solar Nebula are known as asteroids.

Asteroids have been known since 1801, when Ceres was discovered. Since then, astronomers have found more than 300,000 of them. Like Ceres, most orbit the Sun between Mars and Jupiter, forming a doughnut-shaped cosmic junk pile called the asteroid belt. Ceres is by far the largest, at around 950km across – the only asteroid large enough for its gravity to pull it into a sphere. Only three others are larger than 500km. All of these are irregularly shaped, like giant potatoes. Astronomers estimate that there could be as many as 25 million asteroids larger than 100m in diameter, fewer than 0.5 per cent of which are larger than 1km.

Asteroids come in various classes, depending on composition, but in general they are either carbon-rich (the most common), stony or metallic. We know what they are made of because fragments of asteroids – chipped off during collisions – fall to Earth as meteorites. Using sophisticated dating techniques based on radioactive decay, scientists have measured the ages of these samples to high precision, leaving no doubt that they are remnants from the formation of the Solar System, billions of years ago.

It might seem puzzling that the asteroids did not coalesce into a single planet. When asteroids venture too close to mighty Jupiter, on the outer edge of the asteroid belt, they experience a gravitational tug. These asteroids then develop wildly elliptical orbits. Some are flung out of the Solar System altogether, while others plunge into the Sun. This constant gravitational bullying ensures that asteroids will remain, probably forever, individuals – fossilized relics of the original Solar Nebula.

# DWARFS
## SMALL WORLDS

The most recent additions to the Solar System's interplanetary zoo, the dwarf planets, owe their definition to a world called Pluto. Pluto had been designated a planet since its discovery in the 1930s, but there was a problem. A growing number of mysterious bodies on the outskirts of the Solar System – trans-Neptunian objects – were being discovered. Some were almost the size of Pluto. One, Eris, was even suspected (and is now known) to be larger than Pluto itself. Surely not all could be planets? What, in fact, is a planet? Up until 2006 there was no formal definition.

PLANETS AND RELATED SPECIMENS

*We stand on the surface of the icy dwarf planet Pluto. Its largest satellite, Charon, hovers in the sky, cycling endlessly through its phases just like our own Moon.*

In August that year, a group of astronomers met to pen an official definition. A planet, they concluded – though the description is still debated among many astronomers – is an object that orbits the Sun; is massive enough that its gravity makes it spherical; and whose orbit is clear of other objects. The largest known asteroid, Ceres, is not a planet, despite its round shape, because its orbit is shared with other worldlets in the asteroid belt. Similarly, the likes of Pluto and Eris share their orbits beyond Neptune with thousands of other icy worlds larger than 100km across, and trillions smaller still. Pluto, Eris and Ceres are now known as dwarf planets. Two others, Haumea and Makemake, bring the official dwarf planet total to five.

Little is known about these enigmatic worlds, owing to their vast distances – some out to 70 AU. All, except possibly Ceres, are largely composed of ices. Astronomers suspect that around 100 of the known trans-Neptunian objects are also dwarf planets, but being so far away and small makes it difficult to accurately measure their shape.

Like asteroids, the trans-Neptunian objects are probably the remains of the formation of the Solar System. Owing to the vastness of this region far from the Sun, these worldlets collided relatively infrequently, and so their growth into fully fledged planets was curtailed.

"Little is known about these enigmatic worlds, owing to their vast distances..."

If a probe could enter the rings of Saturn and avoid a collision, this is the scene it might record. Countless icy boulders encircle the gas giant, slaves to its powerful gravity. Most are small, but some are as large as a house.

*Scan the page to view animation*

# SATELLITES
## MOONS AND RINGS

So far, everything we have looked at in the Solar System revolves around the Sun. However, the Moon orbits the Earth; it is our planet's only natural satellite. Most other planets have moons as well, as do many asteroids and dwarf planets. Indeed, Jupiter is like a miniature Solar System, boasting at least 67 satellites. Four of them – Io, Ganymede, Callisto and Europa – are substantial in size and spherical. They are called the Galilean moons, after their discoverer, Galileo. Each moon is a unique world, as rich in geology as any planet. They are probably primordial: that is, they formed around Jupiter – in a dusty disc like the Solar Nebula, but smaller – while the planet itself was growing. The smaller moons are probably asteroids ensnared by Jupiter's mighty gravity.

Some of Saturn's and Uranus's moons are large, spherical and primordial, while others are small, irregular gravitational captives. Neptune is the only giant planet that has just captive moons. Of the terrestrial planets, only the Earth and Mars have satellites. Mars's two are irregular – odd-shaped rocks probably captured from the nearby asteroid belt. Earth's moon is a special case. It likely formed when a protoplanet crashed into the newly formed Earth, creating a ring of debris around the young planet, which later coalesced into its satellite.

All the giant planets have rings, the most famous being Saturn's. They are composed of immeasurable numbers of fragments, each like a mini satellite. These particles are so numerous that seen from afar the rings appear solid. Saturn's ring particles are made of ice, varying in size from microscopic to a few metres across. Uranus's rings are made of carbon-rich particles in the range of 0.2m to 20m across. Neptune and Jupiter probably have microscopic ring particles.

*Enceladus is an icy moon of Saturn with a neat party trick – ice geysers. They can spew fountains of icy chemicals, such as propane, ethane and acetylene, 500km into space.*

Up close, a comet resembles an icy asteroid a few kilometres across, spewing geysers of ice from its surface as the Sun heats it up.

Comet tails can stretch for tens of millions of kilometres across interplanetary space. Here, a comet passes behind the Earth and Moon, clearly illustrating the vast extent of its tail.

# COMETS
## AND COMET CLOUDS

We have come to the final class of objects in the Sun's family: the comets. Perhaps the most famous is Halley's Comet, named after the astronomer who correctly calculated its orbit. He examined historical sightings and realized that what had been assumed to be several comets was in fact one object returning every 76 years. Comets, he proved, are like planets – they orbit the Sun with strict periodicity.

Like asteroids, comets are relics of the Solar Nebula, but their compositions are different. Asteroids formed close to the Sun, so they are stony or metallic, whereas comets were fashioned further out, where ices dominated. Probes have shown comets to be 'dirty snowballs', loosely packed chunks of ice mixed with grit and riddled with holes. They measure a few hundred metres to a few tens of kilometres across – much smaller than the largest asteroids.

When far from the Sun, comets are dark and difficult to spot. But because their orbits are highly elliptical, they can approach the Sun closely. Slowly the surface ices start to boil, enshrouding the comet in a giant cloud called a coma. Now two forces come into play. The first is radiation pressure – the push exerted by sunlight itself. The second is the solar wind – subatomic particles blustering outwards from the Sun. These forces push the coma gases deep into space to form a long, glowing tail downstream of the Sun – the hallmark of a comet. Some tails can stretch for tens of millions of kilometres.

Some comets, like Halley, have short orbits, circling the Sun every few dozen years. Others take thousands of years, spending much of their life at great distances from the Solar System hub. To astronomers, this suggests that long-period comets hail from a vast, spherical reservoir, known as the Oort cloud, that surrounds the Solar System. It is perhaps 100,000 AU across. If it exists, as seems likely, the Oort cloud could be home to trillions of inactive comets.

# 9 EXOPLANETS
## NEW WORLDS

The Solar System is a complex environment, with far more to it than a group of planets clinging to a yellow star. Yet that star, the Sun, is but one of billions in our Galaxy alone. Gazing at the celestial dome on a clear night, it is natural to wonder if, huddled around the fires of those countless other stars, strange new worlds exist. The answer is yes – in huge numbers.

The first unequivocal finding of an 'extrasolar' planet (or exoplanet) was announced in 1992, orbiting a dead star called a pulsar (*see* pp50–51). However, in 1995, a discovery really excited the scientific community, when researchers identified a planet orbiting a Sun-like star called 51 Pegasi. Exoplanets are difficult to spot. Next to the light of their parent suns, they are like flecks of dust caught in a floodlight's glare. But 51 Pegasi's planet was found indirectly. As the planet orbits, its gravity tugs on its star, which then moves in a small circle. From Earth, the star appears to wobble left and right. Astronomers can detect this motion and thus infer the presence of the planet.

51 Pegasi is a gas giant, about half the mass of Jupiter. Since its discovery, the number of known exoplanets has exploded – currently, more than 850 are catalogued. When this planetary gold rush began, all the exoplanets were gas giants, but more sensitive instruments now detect smaller worlds. Indeed, a new technique is finding probable terrestrial planets only a few times more massive than the Earth. Each of these planets causes its star's light to dip very slightly (just 1/100th of one per cent) when the planet passes in front of it. This tiny light reduction betrays the presence of a new world. Through this dynamic, important field in astronomy, we now know that planets are the norm rather than the exception. In our Galaxy alone, there could be trillions of alien planets.

"...it is natural to wonder if, huddled around the fires of those countless other stars, strange new worlds exist."

A planet somewhat like Earth encircles a star with a distinctly non-solar hue. This is not our Sun. Sights such as this could be common throughout the Galaxy.

Swollen to monstrous dimensions,
a red giant star sterilizes the surface
of this once verdant world. Billions of
years from now, this could be the view
from the surface of our own planet.

# INTRODUCTION

On a clear night, away from bright city lights, the stars twinkle as we gaze to the heavens. If we could venture among them and look back towards the Solar System, our Sun would all but vanish, lost amid the thousands of other pinpricks of light that surround it. For the Sun is also a star; and the stars are suns, seen from very far away.

Stars are shining spheres of super-compressed, very hot gas. As such, like gas planets, they have no solid surface. A keen observer might notice that stars have subtle tints. This is because they come in a wide range of temperatures. The cooler ones appear red and have 'surface' temperatures of around 2,500°C. Orange or yellow stars, like the Sun, are a little warmer. The hottest stars, with temperatures in excess of 30,000°C, are blue. Stars also exhibit a mind-boggling range of sizes. For example, the gargantuan gas bags known as red supergiants (*see* pages 46–47) measure hundreds of times the diameter of our humble Sun. At the other end of the scale, tiny stellar corpses known as neutron stars (*see* pages 50–51) are no larger than a city – as small compared to our Sun as an ant is to a skyscraper.

Astronomers make sense of all this by categorizing stars according to their colour, temperature and state of evolution. Stars develop constantly, within lifetimes that vary from millions to trillions of years. This stellar evolution is a result of nuclear reactions deep within a star's core. In our Sun, for example, a nuclear furnace converts a staggering two trillion tonnes of hydrogen into helium every hour. That's the mass of a small mountain consumed every second – for billions of years. This releases energy and an outward pressure that stops the Sun from collapsing under its own weight. In time, the hydrogen supply will begin to run dry, and the Sun will change dramatically as it prepares for death.

This chapter will explain how stars evolve – how they are born, how they age and how they die. Only then will it become clear why such a diverse range of sizes, colours and temperatures typify the members of the stellar zoo.

# PROTOSTARS
## YOUNG STELLAR OBJECTS

"Gradually, it unveils itself to the Universe as a huge, luminous mass, glowing like a dull red ember."

Everything has a prelude – even stars. They are born from vast patches of interstellar fog, light-years across, known as giant molecular clouds, which we met briefly in chapter 1. To recap, these nebulae are very patchy, with some regions denser than others. A denser neighbourhood, by virtue of its higher gravity, pulls in material from its surroundings and grows denser still. This increases its gravity even further, which, in turn, allows it to grow bigger and faster. At the same time, rotation within the growing cloud fragment causes it to flatten out. The result, around a million years later, is a gigantic dusty shell called a globule. At the centre of this cocoon is a warming, gaseous clump, about the size of the current orbit of Mercury, known as a protostar.

As its name suggests, a protostar is not yet a true star. While the latter is driven by nuclear reactions, a protostar is powered only by gravitational contraction. To understand this process, picture a dam full of water. Contained, the water has gravitational potential – energy that it has by virtue of its influence under Earth's gravity. If the dam bursts, the water gushes out and its potential is unleashed as kinetic energy, or energy of motion. In the same way, a protostar is full of potential energy. As it falls in on itself under gravity, its potential energy decreases while its kinetic energy, in the form of heat, rises. The contracting protostar gets warmer and warmer.

The protostar stage does not last long. After about 100,000 years, the object becomes warm enough for the heat it generates to start to vaporize its birth cocoon. Gradually, it unveils itself to the Universe as a huge, luminous mass, glowing like a dull red ember. The gravitational contraction continues until, around a million years later, the protostar's core temperature nears five million degrees Celsius. As this prelude to the stellar object's life as a true star draws to a close, it is set to enter a period of turmoil.

If we could peer inside the dark cocoon that surrounds a typical protostar, we might see something like this. The protostar, glowing red, may be flattened out by rapid rotation. It is still surrounded by a vast disc of gas and dust, around which planets will one day form.

A T-Tauri star is a glowing ball of seething energy. It is covered in vast, continent-sized starspots where powerful magnetic fields lower the local temperature and hinder the flow of gases and the transfer of heat. Streams of gas are sucked out of the surrounding disc and plough into the star.

# 2.2 T-TAURI STARS
## STELLAR ADOLESCENTS

A star's T-Tauri phase is violent. The star is still undergoing gravitational contraction, but is not yet hot enough to drive thermonuclear reactions. However, with sufficiently hot surface temperatures of around 4,000°C, virtually all its gases are electrically charged. In a gas, atoms are usually electrically neutral because they are made up of equal quantities of positive and negative charges. But if a gas is hot enough, the heat shakes the atoms apart. This is known as ionization, and the charged atomic fragments are called ions.

So, a T-Tauri star, like a true star, is a gigantic ball of electrically charged gas – a plasma. Compared to a star like the Sun, the average T-Tauri is typically a few times larger and spins more quickly on its axis – once in a few days (the Sun takes a month). Since a moving charge constitutes an electric current, the star's interior is full of electricity. Just as a wire carrying an electric current generates its own magnetic field, the circulating ions inside a star create a global magnetic field. In the case of a T-Tauri star, the rapid rotation creates a much stronger magnetic field than in more normal stars. Furthermore, a T-Tauri star is usually surrounded by the protoplanetary disc that built up around it as it grew. As the star rotates, its magnetic field lines are dragged through this disc. Where the field and disc interact, streams of gas are sucked out of the disc and pulled along the field lines onto the star's surface. Where these packets of gas hit, the star responds with violent magnetic flares – a hallmark of the T-Tauri phase of star formation.

A final characteristic of a T-Tauri star is a powerful stellar wind – a stream of ions that blusters away from the star, deep into space. The Sun has a similar wind, but it is much slower and more rarefied than the one in its T-Tauri adolescence.

"...streams of gas are sucked out of the disc... Where these packets of gas hit, the star responds with violent magnetic flares..."

The Sun is a main-sequence star with a surface temperature of around 6,000 °C. The spots are typical of stars such as the Sun, but these darker regions of slightly lower temperatures are much smaller than those of the T-Tauri phase.

# MAIN SEQUENCE
## NUCLEAR POWERHOUSES

The length of a star's T-Tauri phase varies enormously, depending on the star's mass. For a star like the Sun, for example, this troubled period of stellar adolescence lasts about 30 million years. But in all cases there is a common event – one that ends the T-Tauri stage, stops the star's gravitational contraction and puts it squarely on what astronomers call the main sequence. That event is the ignition of thermonuclear reactions in the star's core, which occurs once the internal thermometer clocks up around 10 to 15 million degrees Celsius.

At these extreme temperatures, the positive hydrogen ions deep within the star's core move so quickly that when they hit, they overcome their mutual electrostatic repulsion and stick together. A series of chain reactions is set up and the hydrogen ions are consumed slowly and converted into a new element – helium. This process, known as stellar nucleosynthesis, releases an enormous amount of energy and a corresponding outward pressure. For the first time in the star's existence, the inexorable inward pull of gravity – which had continuously shrunk the star until now – is halted, exactly balanced by the core's new-found outward pressure. Astronomers call this situation hydrostatic equilibrium, and it is the very definition of the main-sequence phase.

The main sequence is the longest period in a star's lifetime. The Sun has burned hydrogen like this for 4,600 million years, and it is only about halfway to its end. However, as with the T-Tauri period, the duration of the main sequence depends critically on the star's mass. In a much more massive star, the core has so much weight pushing on it that it has to consume its hydrogen supply far more quickly to generate enough outward force to counteract gravity. Such stars tear through their hydrogen stockpiles in only a few million years. But very low-mass stars, staggeringly, live a million times longer. They are known as red dwarfs.

**"For the first time in the star's existence, the inexorable inward pull of gravity ... is halted"**

# RED DWARFS
## FEEBLE AND FAINT

Main-sequence stars are known as dwarfs, which does not mean they are small – it is simply that more evolved stars tend to be bigger. Red dwarf stars, however, are deserving of the term, having only a fraction of the Sun's mass, diameter and luminosity. So dim are they that not one is visible to the naked eye – the nearest, Proxima Centauri, is 100 times too faint to spot.

Red dwarfs have such a puny mass that the pressure at their centres – from the weight of the rest of the star – is relatively low. Ergo, these stars do not need to generate much energy to prevent gravitational collapse, so they consume their hydrogen stockpiles sedately. In addition, red dwarf interiors are fully or mostly convective, like pans of boiling water. The heat generated in the core rises to the surface where it spreads out, cools and sinks down to be reheated. Basically, red dwarfs are like huge convection heaters, with thoroughly churned and mixed interior gases. In a Sun-like star, by contrast, the gases higher up do not mingle with those in the core. Once a Sun-like star exhausts its core hydrogen, the rest of the hydrogen higher up remains largely untouched, like a campfire that burns in the middle and leaves the edges unscathed. Inside a red dwarf, even hydrogen on the surface is transported to the core, where it is consumed during the main sequence. This combination of a slow fusion rate and a large hydrogen supply means that a star like Proxima is expected to be on the main sequence for an astonishing four trillion years – 290 times the age of the Universe!

Still, not even the stars are eternal. Once the hydrogen supply begins to run dry, replaced slowly and inexorably by a helium 'ash', the stage is set for the next phase in a star's evolution. Leaving the main sequence behind, it becomes a subgiant.

"So dim are they that not one is visible to the naked eye – the nearest, Proxima Centauri, is 100 times too faint to spot."

Red dwarfs are the most common stars in the known cosmos. Of the 50 closest stars to the Sun, 40 are red dwarfs. This image shows a typical example, feeble and dim, with a volcanic planet in close attendance.

# SUBGIANTS
## GROWING OLD

Once a star begins to run out of the hydrogen gas it needs to generate its life-sustaining nuclear reactions, it is on dangerous turf. Its long period of stability on the main sequence ends – and death is near. The problem is helium – the product of a star's hydrogen reactions. Helium ions carry two times the electrical charge of their simpler hydrogen counterparts. In order to fuse helium ions in a nuclear reaction, they have to move much quicker than hydrogen ions to overcome their mutual electrostatic repulsions and bond. In other words, much higher temperatures are needed to fuse helium than to fuse hydrogen. So when the hydrogen runs dry, the nuclear reactions in the star's centre turn off, and gravity takes over. The star's core, lacking its internal pressure source, starts to shrink.

However, this renewed contraction brings a new, brief respite. While the star was on the main sequence, it depleted the hydrogen in its core but not in those regions above it (unless the star is a red dwarf). The gravitational squeeze, therefore, brings this fresh, unused fuel into the core region, where it is hot enough to react. Once again, the star is stable, converting hydrogen into helium in a narrow shell on the outskirts of the dead, reactionless helium core. As a result of this new period of thermonuclear frenzy, the star's outer layers expand with the renewed energy pouring into them. While the core continues to shrink, the star as a whole grows gradually larger and cooler. At this stage, astronomers call the star a subgiant.

Subgiants constantly evolve and expand, while their cores contract and heat up. For a star like the Sun, this phase is expected to persist for 700 million years, with the star swelling to three times its main-sequence diameter by the end. Eventually, though, even the hydrogen shell is exhausted and this triggers another crisis. The star is then poised for its most dramatic metamorphosis to date.

There is little to tell a subgiant star apart from a younger one on the main sequence. Altair, a white star in the constellation of Aquila, or the Eagle, is thought to be a subgiant, but astronomers are not entirely sure.

"Its long period of stability on the main sequence ends – and death is near."

# GIANTS
## GERIATRIC STARS

Subgiants may be dramatic, but they are puny compared to the stars that they are destined to become. A subgiant's core is contracting all the time, and thus heating up as it does so. Once the core temperature is high enough, so much energy is generated that the radiation pushes on the star's outer layers and puffs them up to even greater dimensions. The star swells into a monstrous caricature of its former self. With a typical diameter of a few dozen to hundreds of times larger than a normal main-sequence star, and 1,000 to 10,000 times brighter, these stars are known as red giants.

When the Sun reaches the end of its life, it will expand into a red giant, with an expected diameter of around 200 times its current size – easily large enough to engulf the innermost planets, possibly including the Earth. Eventually, with its core constantly shrinking as it runs out of reactionable fuel, its internal temperature will reach a staggering 100 million degrees Celsius. That's the magic number needed to fuse helium ions together. Suddenly, nuclear reactions begin anew, this time feeding on the helium and creating a new ash – carbon and oxygen. The core, alive once again, spews fresh energy outwards, puffing out the star's outer layers even more, until they start to escape from the star altogether, forming a planetary nebula (*see* pp106–107).

Generally, astronomers use 'giant' to describe stars that are much larger than main-sequence stars of the same temperature and hence colour. So, not all giant stars are red. Alcyone, a star in the constellation of Taurus, is a blue giant that has exhausted the hydrogen in its core and has swollen beyond its main-sequence radius. It's on the way to becoming a red giant, as are most other non-red giant stars. For many stars, including the Sun, a red giant is the end point of stellar evolution. However, for the most massive stars, heavier than about eight to ten Suns, the end is even more impressive.

> **"When the Sun reaches the end of its life, it will expand into a red giant, with an expected diameter of around 200 times its current size…"**

A blue giant star, powerful and ultra-luminous, ravages the surface of this nearby rocky world. These stars can pump out 10,000 times as much energy as the Sun.

# SUPERGIANTS
## GARGANTUAN GAS BAGS

The most massive stars, outweighing the Sun by factors of eight or more, do not follow the same evolutionary trend – main sequence, subgiant, giant and red giant – as their smaller siblings. Rather, they evolve off the main sequence into exceptionally luminous powerhouses called blue supergiants. These monsters are unimaginably bright, some shining with the brilliance of at least 10,000 Sun-like stars – and in some cases, up to a million.

Despite this power and their great mass, blue supergiants are not as large as their name suggests. The biggest are around 25 times the diameter of the Sun. This is certainly huge, but the red giants we met earlier are an order of magnitude bigger still. The difficulty is that, in astronomy, terms such as 'giant' and 'supergiant' are used – somewhat counterintuitively – not to refer to the star's size but to its state of evolution. Younger stars (main sequence) are dwarfs; more evolved ones are subgiants; and the most evolved are red giants. A very massive, bright main-sequence star might be bigger than a more evolved but less massive subgiant.

The largest stars, though, are the red supergiants, which evolved from blue supergiants. Blue supergiants are so powerful that they constantly lose gas – they are literally blown apart by the pressure of their own intense radiation. This puffs them up to dimensions that give new meaning to the word 'huge'. While red giants – formed from relatively low-mass stars – may approach 200 solar diameters, the largest known red supergiant, VY Canis Majoris, is an astonishing 2,000 times the size of our Sun. If placed at the centre of the Solar System, it would engulf all the planets out to and including Saturn.

Some stars evolve from blue supergiants to red supergiants, then back to blue again. This happens because as the red star loses its outermost layers, the hotter, bluer and smaller interior is revealed. The star settles down again as a blue supergiant, before billowing up once more to its supersized, red alter ego.

Red giants are not perfect spheres. Because they are convective, their envelopes are made up of bubbling cells of gas, which rise to the surface and spread out before sinking. In this image, a luckless planet is destroyed as its star expands to truly monstrous proportions.

# WHITE DWARFS
## STELLAR CORPSES

The stars we have met so far have been 'alive' – in the sense that they hold their own against the crush of gravity by fusing elements together deep within their cores. But when this nucleosynthesis ends, gravity finally wins. The star faces death. A low-mass star, as we have seen, becomes a red giant and sheds its outer layers to form a planetary nebula (see pp106–107). When this process is complete, all that is left of the once mighty star is its naked core – inert, yet still very hot: a white dwarf.

White dwarf stars are made of helium, or carbon and oxygen – the products of nucleosynthesis. They are exceedingly compact. Once a star leaves the main sequence, its existence is a constant battle as the core struggles to generate enough energy to balance gravity. For hundreds of millions of years (in the case of a solar-mass star), the core contracts. All its material will eventually be squeezed into a super-dense sphere with only one-millionth of the star's original volume. The result: a typical white dwarf no larger than a terrestrial planet like Earth, though it may outweigh the planet by a factor of 200,000. White dwarfs, therefore, have absurdly high densities. Just a teaspoon of their substance would weigh as much as a pickup truck. A material such as this is called a degenerate gas. Essentially, the electrons are crammed so close together that there is no space between them. So, while white dwarfs are inert, they are stable. Not even gravity can squeeze the electrons closer together.

Because white dwarfs lack an internal energy source, they resemble planet-sized lumps of coal. Gradually, they are destined to grow cooler and darker until they emit no light at all. They are then called black dwarfs. However, astronomers have a long time to wait before they detect any, for a black dwarf is expected to take several billion times the age of the Universe to form – a mere ten million trillion years!

**"All that is left of the once mighty star is its naked core – inert, yet still very hot: a white dwarf."**

The shrill glare of a white dwarf spills over the battered landscape of a world once full of life. A fate such as this awaits Earth in the distant future – if it is not swallowed first when the Sun becomes a red giant.

WHITE DWARFS

A pulsar is a city-sized sphere of super-dense matter, emitting beams of light and spinning sometimes hundreds of times per second. If the pulses intercept the Earth, we observe a cosmic lighthouse as the radiation beams sweep across our line of sight.

# 2.9 NEUTRON STARS
## THE ULTIMATE DENSITY

We have seen what happens to a star of low to moderate mass when it dies. In the case of a heavier star, the final death throes are much more energetic (*see* pp108–109) and the corpse is an object even stranger and denser than a white dwarf – a neutron star.

White dwarfs are balanced against collapse by electron degeneracy pressure. For stellar cores more massive than about 1.4 solar masses, however, even this is not enough. The relentless gravitational crush causes the ions of helium, oxygen or carbon to shatter into their constituent subatomic particles, called protons and neutrons. The protons join with electrons – positively and negatively charged, respectively – to make more neutrons. Suddenly, the core is full of neutrons and little else. This frees an enormous amount of space and allows the core to contract even further, stopping only when the neutrons reach degeneracy. Neutron stars are far denser than white dwarfs; as dense as the entire human population, seven billion people, squashed into the volume of a sugar cube. These stars are no larger than a small city, mere kilometres across.

Another astonishing quality of neutron stars is their rapid rate of spin. Just as an ice skater spins up when she pulls in her arms, a star spins quicker as it shrinks. Since a neutron star is around 50,000 times smaller than its main-sequence radius, the increase in rotation is phenomenal. These stars can spin on their axes hundreds of times every second. In addition, they have super-amplified magnetic fields, ten billion times stronger than a fridge magnet. These fields cause certain neutron stars to emit beams of light and radio energy along their magnetic axes. If the Earth is ever in sight of these beams, we receive a pulse of radiation with each turn of the star, hundreds of times per second. These neutron stars are known as pulsars.

> "...we receive a pulse of radiation with each turn of the star, hundreds of times per second."

# BLACK HOLES
## LIGHT TRAPS

White dwarfs and neutron stars are certainly awesome, but a stranger fate awaits the Universe's heaviest stars. White dwarfs are stable if they are less massive than about 1.4 solar masses. Any larger and gravity overwhelms the pressure of the degenerate gas and the ions shatter, forming a neutron star. Neutrons are crammed against each other; the star is essentially a city-sized subatomic particle. These stars withstand their own gravitation because there is no space between the neutrons – but only up to a point. In neutron stars weighing more than three solar masses, the neutrons splinter and then nothing known to science can stop the star collapsing indefinitely. It shrinks until it has zero size and infinite density. Only its incredible gravity remains. These exotic stellar corpses are called black holes.

On Earth, when you throw a ball up, gravity slows it down; it reverses direction and falls to the ground. If a ball travelled fast enough, though (at about 11km/s), it would overcome gravity and escape into space. A black hole's gravity is so powerful, however, that no object would ever escape it. Even light – which travels at the Universe's ultimate speed limit of 300,000km/s – is held down. Black holes emit no radiation, including light, of any kind. This is what gives them their name.

Astronomers have found many cases where an invisible object, a suspected black hole, orbits a normal star. The black hole's gravity warps its companion into an egg shape, pulling gas from its atmosphere. Just before they vanish into that gaping maw, the gases are so hot that they shine brilliantly in X-rays. This is the phenomenon we met earlier, where gravitational potential energy is converted to heat as the gases flow inwards. Astronomers detect this and infer that a black hole must be there. They can measure its mass using the laws of motion, and be confident it is a black hole if it is heavy enough. We will meet some of these binary systems in the following chapter.

This black hole is surrounded by a swirling pancake of gas, stolen from a normal companion star (not shown here). The gases are heated to extreme temperatures and emit vast quantities of deadly X-radiation, betraying the hole's presence.

**Scan the page to view animation**

BLACK HOLES

53

Achernar is a massive blue star spinning at breakneck speed. The rotation flattens the star at the poles and centrifugal forces help create a disc of gas surrounding the star.

# BE STARS
## RAPID ROTATORS

Before we leave this chapter, we will spend a few pages exploring more stellar oddballs. First, let's look at Be stars (pronounced 'B E stars'). These stars are spinning so quickly that they are oblate – or bulging – at the equator.

There are examples of oblate objects closer to home, in our Solar System. All the planets, but especially Jupiter and Saturn, have a shape called an oblate spheroid. They are not truly spherical. Saturn spins so quickly – once every 10.6 Earth days – that its equator is ten per cent wider than its polar diameter. This is because Saturn is a world made of fluid. As it rotates, centrifugal force causes it to swell at the equator. Similarly, since stars are mostly fluid, some of them are oblate as well. The amount of centrifugal bulge can be enormous.

A celebrated example is Achernar, the brightest star in the constellation of Eridanus (the River). Achernar is a hot, blue Be star around seven times the mass and radius of the Sun, and well over 3,000 times its luminosity. It is the most oblate star known, with an equatorial diameter an astonishing 56 per cent greater than its diameter measured pole to pole. Because of this highly flattened shape, the polar regions are hotter than the equator, since they are much closer to the star's nuclear reactive core. So, while the poles bask in temperatures of around 20,000°C, the equatorial zone is only around half as hot.

The rate of spin is so great that many Be stars are surrounded by a vast, thin disc of gas, flung off from the stars as they rotate – in the same way that centrifugal forces eject water from wet clothes in a spin drier. The exact mechanism that creates the disc is uncertain, though, and it is likely that magnetic fields are also involved in its formation.

**"The spin is so great that many Be stars are often surrounded by a vast, tenuous disc of gas, flung off from the stars as they rotate..."**

# FLARE STARS
## STELLAR MAGNETS

Astronomers frequently encounter magnetic fields, which permeate the great gas clouds, or nebulae, throughout interstellar space. Most Solar System planets have magnetic fields and we have already encountered them in association with some stars – T-Tauri stars and neutron stars, for example. In a particularly potent class of neutron stars known, quite fittingly, as magnetars, the field strength is so intense that it could wipe credit cards and rip cutlery out of your hand, from a distance of halfway to the Moon! The so-called flare stars, while not quite as extreme, are no less awesome.

Flare stars are those in which magnetic activity causes the stars to brighten suddenly with brief, violent bursts of energy. Similar events take place on the Sun – solar flares. These occur when magnetic energy accumulates in the Sun's atmosphere and is suddenly released, usually close to the zones around sunspots. Solar flares are epic events, releasing some six septillion joules of energy (that's a six followed by 25 zeros). To give some perspective, a solar flare releases energy equivalent to around 160 billion megatons, or eight billion Hiroshima-strength nuclear warheads. Yet the Sun is a fairly sedate star. In proper flare stars, the magnetic fields are typically a thousand times stronger than the Sun's. During a flare, the star jettisons vast quantities of subatomic particles into space at speeds approaching that of light, and the star's luminosity increases drastically. One example, a star called UV Ceti, emits up to 75 times its normal amount of light during a flare episode.

The flaring phenomenon works only in relatively cool stars – yellow, orange or red – and the overwhelming majority of them are red dwarfs. Indeed, the red dwarf Proxima Centauri is also a flare star. More recently, astronomers have discovered flaring activity in star-like objects called brown dwarfs, which we will meet next.

"...the magnetic fields are typically a thousand times stronger than the Sun's."

In a typical flare star, flaring occurs persistently – though the process is abrupt and unpredictable – with each one lasting minutes or hours. Usually, gigantic spots punctuate the toiling stellar surface of these stars.

# BROWN DWARFS
## FAILED STARS

The final objects in this chapter are not really stars – yet they are not planets, either. They occupy the middle ground between massive planets and flyweight stars. These are the brown dwarfs.

Stars are created when a gas cloud collapses under gravity, forming a central mass with a disc of gas and dust around it. Astronomers think that brown dwarfs are created in the same way. However, they never accumulate enough mass to synthesize helium from hydrogen. The reactions to do this require a core temperature of at least ten million degrees Celsius, but a typical brown dwarf's core never gets hotter than about half that, and often it is a hundred times cooler still. For this reason, some astronomers refer to brown dwarfs as failed stars. Nevertheless, their cores are hot enough for a special form (or isotope) of hydrogen, called deuterium, to undergo fusion reactions, leading to the synthesis of a helium isotope called helium-3. The more massive brown dwarfs are able to produce ordinary helium by consuming another element, lithium. So, despite not making the grade as true stars, the heaviest brown dwarfs do still emit a substantial amount of light, courtesy of their nuclear cores, which can be comparable to that emanating from the faintest stars.

Brown dwarfs vary in size, from around the diameter of Jupiter to a few times larger, which makes them much denser than a planet. Typically they weigh in at between 13 and 75 Jovian masses. They have surface temperatures ranging from the freezing point of water to around 2,500°C. To the naked eye, a brown dwarf appears deep red or magenta, depending on how hot it is, although those with low surface temperatures emit no detectable light at all. Intriguingly, some of these objects have weather patterns – planet-sized storms and swirling cloud bands, similar to those of Jupiter. Unlike Jupiter, though, clouds found in the atmospheres of brown dwarfs are made of hot grains of sand and iron.

**"They occupy the middle ground between massive planets and flyweight stars."**

A brown dwarf is a cosmic oddball – neither a planet nor a star, yet with traits of both. It is enveloped in dark bands of cloud, obscuring the warmer, brighter layers below. In this image, a brown dwarf is orbiting a normal star (left).

# 3 CAMARADERIE OF THE STARS

Two stars, each seething with magnetic energy, scoot around each other in a tight orbit. Called RS Canum Venaticorum, this is just one example of a close binary star.

# INTRODUCTION

Like living organisms, stars favour companionship. The pinpoints of light that speckle the night sky are not all individuals; many are pairs or even higher multiples. They are not simply chance groupings, either – two or more stars that merely appear close. No, these are actual physical associations – stellar systems in which two or more members engage in a stately gravitational dance about their centre of mass. Sometimes, as you will soon learn, this dance turns frenetic.

Astronomers estimate that between one-quarter and one-half of all the stars in our Galaxy are binary (two stars) or multiple systems (more than two). If you examine the 20 stars nearest to the Sun, for example, you will find that there are not 20 stars at all – there are around 30. Two are triple systems (including Alpha Centauri, the closest star system to the Sun) and six are binary. Only the remaining 12 stars are truly single.

How does this stellar camaraderie arise? Multiple stars are common because, in general, that is how stars are born. We know this simply from the sheer scale of interstellar space. Imagine that the Sun was pea-sized (a scale reduction factor of about 140 billion). Even at this small scale, it would be several hundred kilometres before the Sun's nearest neighbours were encountered. Space is mostly empty – it's called 'space' for good reason. The sheer nothingness that pervades the Universe shows that multiple stars are, generally, not the result of stellar neighbours just happening to pass by each other. While chance encounters do occur, they are too rare to account for the degree of binaries and multiples seen. The only conclusion is that stellar multiplicity is a product of stellar formation. In this chapter, we meet some of these gregarious stars and learn about stellar evolution along the way.

We witness a binary star with widely separated blue and red components, and a hypothetical terrestrial planet in attendance. The planet is orbiting the closer red star, which itself revolves around the more distant blue star.

# WIDE BINARIES
## TWO STARS FAR APART

Binary star systems come in a bewildering variety of forms. Sometimes, individuals are extremely close (so much so that they actually interact). These are the semi-detached or overcontact binaries, many kinds of which are featured in this chapter. Usually, though, the individuals are far apart: perhaps as much as an astronomical unit (AU) – defined as the Earth–Sun distance (*see* p17) – or hundreds of times greater. These wide binaries can take decades or even centuries to complete their gravitational ballets. The two stars live separate lives, each orbiting the other but never venturing too close, for perhaps many billions of years.

The chances are you have gazed at a binary star system. About 8.6 light-years away, in the constellation of Canis Major (the Great Dog), Sirius is the sky's brightest star – although it's actually two stars. The largest is Sirius A. It is a little bigger than the Sun, hotter and therefore blue-white in colour. Its smaller partner, Sirius B, is a white dwarf – a stellar corpse with as much mass as our Sun packed into a super-dense sphere just a bit larger than the Earth. Sirius A and B are several AU apart, taking 50.1 years to orbit each other. More accurately, they orbit not each other, but their 'barycentre' – the point between them where they would balance if you could join them with a large enough stick. As they orbit, their elliptical paths bring them alternately closer together and then farther apart.

Imagine living on a planet of a binary star. The glorious sight of a double sunrise would greet you in the morning, and a double sunset would follow at dusk. Gradually, as the stars moved apart, there would come a time when one star rises as the other sinks below the horizon, holding nightfall at bay for several days, or even years, depending on the mechanics of the system.

*On a world not unlike our Earth, a pair of widely separated stars makes a stunning double sunset.*

*A nova explosion rips through the accretion disc surrounding the white dwarf and blasts the shattered disc into space. This forms a glowing shell of gas surrounding both stars out to great distances. The white dwarf, too small to be seen on the scale of this image, is buried in the blue glow adjacent to its larger, red partner.*

# NOVAE
## EXPLOSIVE INTERACTIONS

In some binary systems, the two stars are so close that one reaches out and touches the other. The cataclysmic binaries, or cataclysmic variables, are one such group, so named because the interaction often has profound and devastating effects on one or both stars. Perhaps the most well-known examples are novae (singular, nova). The name, from the Latin word for 'new', originally described brand-new stars that seemed to appear by magic. In reality, a star too faint to be seen suddenly becomes tens of thousands of times brighter than before.

Imagine a system where two stars orbit each other in just a few hours, so tightly that their separation barely amounts to a solar diameter. The larger of the two is a red dwarf: a dim and unassuming main-sequence star. The smaller star is not so docile. It is a white dwarf: a planet-sized stellar corpse. Each pulls on the other with unyielding determination, but they play a gravitational tug-of-war that only one star can win. Giving in to the gravitational demands of its cannibalistic partner, the red dwarf's atmosphere plunges towards the white dwarf – ten thousand million tonnes of gas every second surrounds the compact star in a glowing ring of plasma spanning half a solar radius. Gradually, the ring thickens and spreads into a so-called accretion disc that reaches all the way to the white dwarf's surface – where the gas accumulates.

Eventually, temperatures, pressures and densities on the white dwarf's surface reach extremes. The material becomes dense and hot enough to ignite thermonuclear reactions, as occurs at the heart of a star. Suddenly, a huge explosion rips through the gas, announcing its presence to anyone within hundreds of thousands of light-years. A beacon of a nova is born.

Usually, the event is final. But in some systems the white dwarf accumulates gas some time after the initial outburst, leading to subsequent eruptions. These are recurrent novae. There are perhaps 25 to 50 nova detonations occurring every year in our Galaxy alone.

> "Suddenly, a huge explosion rips through the gas, announcing its presence to anyone within hundreds of thousands of light-years. A beacon of a nova is born."

# DWARF NOVAE
## REPEAT OFFENDERS

A second type of cataclysmic binary, the dwarf novae are a little like sedate versions of their explosive cousins. As before, the stage consists of a close pair of small stars locked together under gravity – a white dwarf and a red dwarf. Once again, an accretion disc develops as the larger but more lightweight red dwarf succumbs to the pull of the hot and powerful white dwarf, a foreboding presence less than a solar diameter away.

A dwarf nova differs from a nova, both in the mechanism that produces the outburst and in its power. In a nova, the cataclysmic eruption – the sudden, explosive and uncontained ignition of thermonuclear reactions on the white dwarf's surface – is devastating. The disc is blown to shreds, blasted into space, and only rarely does it re-form. But in a dwarf nova, the rapid brightening is much less violent and powerful. It derives from a temporary increase in the rate at which material is channelled through the disc towards the white dwarf. Abruptly, with more gas plunging through the disc, it becomes larger, hotter and brighter, its luminosity flaring up by a factor of between ten and a few hundred. That's significant, but it's still far less than in a classic nova. Eventually, the accretion rate diminishes and returns to normal. The disc slips back to its slimline, quiescent state, and the whole cycle begins again. The timescale between successive dwarf nova outbursts varies, as does the duration of the events themselves. Some systems blow their tops every few days, while in others a wait of several decades is not uncommon. The outbursts generally last anything from a couple of days to three weeks.

Although the outbursts in novae and dwarf novae are not the same, dwarf novae systems can and do experience more powerful nova explosions once enough material has been transferred to the white dwarf.

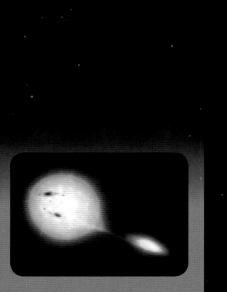

*After the outburst, with the accretion rate reduced, the disc is empty and small. The system can again enjoy its quiescence.*

Just prior to the outburst, the accretion disc is fat and bloated, having gorged itself on star-stuff. The tiny white dwarf, typically several dozen times smaller than its red dwarf partner, is at the centre and is too small to be seen here. The surface of the disc is bright, illuminating the facing surface of the red dwarf.

**Scan the page to view animation**

# DQ HER STARS
## ACCRETION CURTAINS

Cataclysmic binaries are diverse. Novae and dwarf novae are two of this group's protagonists. Now let's meet another. Most solitary white dwarfs harbour a degree of magnetism, with a magnetic field strength that can be tens of thousands of times stronger than a fridge magnet. So, it's reasonable to think that magnetic fields also exist in white dwarfs found in cataclysmic binaries. In the subclass of cataclysmic binaries called DQ Her stars – more commonly known as intermediate polars (IPs) – the white dwarf has a magnetic field up to a few million times stronger than Earth's.

The red and white dwarfs in a typical IP orbit each other every few hours. As they do so, the white dwarf spins rapidly on its axis, with one rotation ranging from just tens of seconds to an hour or so. These characteristics create emissions that can be observed. Because the white dwarf and accretion disc are so hot, often with temperatures up to hundreds of thousands of degrees Celsius, cataclysmic binaries generally emit large amounts of high-energy radiation. They are also brighter at X-ray or ultraviolet wavelengths than in visible wavelengths. In IPs, these emissions vary with the spin period of the white dwarf as well as the binary orbital period.

IPs, unlike non-magnetic cataclysmic binaries, do not have normal accretion discs. If the orbital period is fairly long – say more than four or five hours – or if the magnetic field is less than a certain value, an accretion disc will likely grow as usual, but have its centre disrupted by magnetic forces. Gas from the inner edge of the resulting narrow ring attaches itself to the magnetic field and plummets at supersonic velocities towards the white dwarf. When it hits the star's surface, its kinetic energy converts to radiant energy and is blasted away as X-rays. If the magnetic field's strength is much higher, or the binary separation is slightly tighter, the magnetic field prevents a disc from forming at all. IPs without discs share this distinction with another magnetic class of cataclysmic binary, which we will meet next.

The white dwarf's magnetic field in an IP binary is so strong that it acts on the accretion disc like a whisk in cake mixture, sweeping out the gas in the centre. The white dwarf feeds on this shortened disc, pulling in great curtains of gas from the ring's inner edge and drawing it towards its magnetic poles. This structure is known as an accretion curtain.

# POLARS
## SUPER MAGNETISM

Not long after astronomers unearthed the secrets of intermediate polars (IPs), a star system in Hercules started to attract attention. In 1977, researchers found another magnetic binary, but this one had a startling twist: it seemed to possess a phenomenal magnetic field strength 30 times greater than in IPs, a full 100 million times the strength of Earth's. Since then, dozens of systems have been discovered with similarly strong magnetic fields. These are the AM Herculis stars – also called polars because their strong magnetic fields render their light highly polarized. They form a fourth significant subclass of cataclysmic binary, after novae, dwarf novae and IPs.

Polars have even more profound behaviour than IPs. Gas slipping away from the red dwarf's atmosphere is seized by the powerful magnetic field soon after it becomes free. It cannot circle the white dwarf to form an accretion disc. Instead, the gas flows like a gigantic auroral arc from the surface of the red star along magnetic field lines to the white dwarf's magnetic poles. Imagine that the red dwarf is a ball of molasses and picture the magnetic field as loops of wire. As those wires rotate, they are dragged through the molasses and begin to get stuck. This slows down the rotation of the ball of molasses until, eventually, it stops completely. What results is a binary system where the white dwarf's magnetic field becomes permanently lodged in the atmosphere of its companion, So as the white dwarf spins, the red dwarf rotates with it. In the same way that the Moon keeps the same face towards Earth, each of the two stars keeps the same face towards the other, and the whole system of stars and flowing gas essentially rotates as a solid body. Moreover, because the white dwarf has stopped rotating relative to its companion, the radiation from polars varies only with the orbital period – there is no separate spin period.

> "...the gas flows like a gigantic auroral arc from the surface of the red star along magnetic field lines to the white dwarf's magnetic poles."

*This polar, known as SDSS-1212, differs from other polars in that a brown dwarf (left) replaces the usual red dwarf. The brown dwarf's atmosphere slips away, flowing in a narrow stream towards the white dwarf (right) in close attendance. The flow is fairly straight, but as it approaches the compact star, it latches on to its magnetic field and flows towards the magnetic poles.*

Let's move away from cataclysmic binaries to ponder an even more energetic class of objects: powerful X-ray sources known as X-ray binaries. In an X-ray binary, the compact star or primary – the one feeding off its larger partner – is a neutron star or, sometimes, a black hole.

Neutron stars and black holes, like white dwarfs, are stellar carcasses left over from the deaths of initially much larger stars. But neutron stars are around 100 times smaller than white dwarfs and black holes are punier still, making X-ray binaries exceptionally bright and powerful. To see why, imagine a white dwarf at the centre of an accretion disc. As the gas spirals around the star, the gas gathers momentum and gets hotter – its kinetic energy increases. When it finally hits the white dwarf, the gas halts abruptly and its kinetic energy converts into light, ultraviolet radiation and sometimes X-rays. Now imagine an X-ray binary. Instead of a white dwarf at the centre of the disc, a neutron star (or a black hole) lurks. Because these are so much smaller than white dwarfs, the material in the disc can spiral much further inwards, picking up much more speed and heat before it encounters the neutron star or the black hole. As a result, the energy liberated in an X-ray binary is truly phenomenal. A typical system emits as much energy in X-rays alone as hundreds or even thousands of Sun-like stars combined.

X-ray binaries come in two varieties depending on the mass-donating or secondary star. If the secondary is lightweight, say comparable to the Sun or smaller, then the system is called a low-mass X-ray binary (LMXB). Often, the secondary in an LMXB is a red dwarf. However, if the secondary is massive, much greater in mass and size compared to the Sun, then it is a high-mass X-ray binary (HMXB), which is dealt with overleaf.

**"A typical system emits as much energy in X-rays alone as hundreds or even thousands of Sun-like stars combined."**

This is an *LMXB* as seen from a hypothetical rocky world nearby, perhaps a drifting asteroid. Because the secondary star is small and hugs its compact companion closely, some *LMXBs* could easily be contained entirely within our Sun. Visually, they look a lot like cataclysmic binaries.

# HMXBs
## HIGH-MASS X-RAY BINARIES

High-mass X-ray binaries (HMXBs) easily dwarf their puny, low-mass cousins. These systems typically comprise a star – at least ten times more massive than the Sun – that orbits a neutron star or black hole.

Often, the secondary star in an HMXB is a blue giant: a very powerful, massive star with a strong stellar wind – a steady stream of subatomic particles blustering away from the star's atmosphere and deep into space. Although the giant stars in some HMXBs may lose material to their greedy partners via the accretion disc, this does not always occur. Indeed, often there is no disc at all, for the black hole or neutron star may be too far from its cohort to strip it of gas. Nevertheless, mass transfer still takes place, courtesy of the giant star's stellar wind. Flowing into space, radially outwards in all directions, some of the stellar wind encounters the neutron star or the black hole's gaping maw. There, like water swirling around a stone in a stream, the flowing gas wraps around the compact object and a wake forms behind it – a bow shock. As the gas is brought to an abrupt halt at the bow shock, the compact star snatches it and pulls it in, vigorously generating X-rays.

A particularly well-known example of an HMXB is Cygnus X-1, because it was the first virtually sure detection of a black hole. Black holes can never be seen – they emit no light. But the radiation emitted by an X-ray binary from the superheated accretion disc that surrounds it means that something tiny and dense must be orbiting and devouring the giant star. If it is only a bit weightier than the Sun, it is a neutron star. If it is much more massive, it can only be a black hole. In this way, astronomers are as certain as they can be that a black hole lurks at the hearts of at least some LMXBs and HMXBs such as Cygnus X-1.

A blue giant (right) orbits a tiny, dense companion, either a neutron star or a black hole. The latter is buried in the midst of the bow-shaped flow of plasma that surrounds it.

In this depiction of a microquasar X-ray binary, jets of subatomic particles, moving almost at a quarter of light speed, spew from the centre of the accretion disc and head into deep space.

Scan the page to view animation

Before leaving X-ray binaries, there is one more class of objects we need to meet: the microquasars. About 17,000 light-years away in the constellation of Aquila lies a curious X-ray source known as SS433. An X-ray binary, SS433 is conspicuously bright in visible light as well as in X–rays, making it an easy target for astronomical research. It emerged early on that SS433 was unlike any other X-ray binary. What makes it different are the high-energy jets of gas that emanate from the system and plunge into deep space at almost a quarter of the speed of light. They are seen in optical, radio and X-ray bands.

In 1992, astronomers found a pair of jet powerhouses with similarities to SS433. Two more were discovered in 1994. Now, SS433 is one of a handful of X-ray binaries all emitting fast particle jets, dubbed microquasars. They were named for their similarities to quasars (*see* pp128–129). While quasars involve the cores of active galaxies, billions of light-years away, where accretion discs surrounding supermassive black holes emit similar jets on far larger scales, the microquasars are stellar in origin – and much closer to home.

In one system, nicknamed 'old faithful' after the famous geyser in North America's Yellowstone National Park, the X-rays from the source wink off every half an hour, prompting the sudden appearance of radio and infrared jets. Five minutes later, the jets vanish and the X-rays reappear. A possible explanation for this odd activity is that in the system – which goes by the prosaic name of GRS 1915+105 – the accretion disc periodically fills up with material stolen from the bloated secondary star by the black hole. During this time, the disc is a strong X-ray source due to the power of accretion. However, once the disc is full, after about 30 minutes, the system blasts the inner portion of the disc into space. At this point, the X-rays rapidly diminish and the ejected material forms the radio and infrared jets.

> "...high-energy jets of gas that emanate from the system and plunge into deep space at almost a quarter of the speed of light..."

# 3.9 RS CVN STARS
## A MAGNETIC EMBRACE

The constellation of Canes Venatici, representing two hunting dogs in the northern sky, boasts an intriguing binary star – a prototype of an entire class of similar objects called RS Canum Venaticorum (RS CVn) systems. The two stars in an RS CVn system are known to be binary because of the behaviour of their spectrum – the pattern that emerges when a star's light is split into a broad strip of colours – and because, like Algol stars (*see* pp82–83), the stars periodically eclipse each other as they orbit.

Now, because the orbital period is so short, these stars spin at breakneck speeds in that same time frame (hours or days). The Sun, by comparison, takes the best part of a month to complete one rotation. Since stars are largely ionized – that is, the atoms in their gases have split into positively and negatively charged ions – a spinning star is essentially a giant dynamo. As it turns, its ions create currents in the stellar interior that in turn generate a global magnetic field. The rapid rotation in RS CVn systems is what makes them conspicuous, because the faster the spin, the more powerful a magnet a star becomes.

Just as the Sun's magnetism creates localized cool regions called sunspots, this stellar acne can afflict other stars, too. In RS CVn stars, the sunspots are vast – patches that cover huge fractions of their troubled stellar surfaces.

A typical RS CVn binary consists of two stars in a tight gravitational bind, only a few stellar radii apart. Consequently, the orbital period is merely several hours to a few days. One star is always an orange subgiant; the other is somewhat like the Sun, yellow or occasionally a hotter yellow-white. As these stars orbit each other, each keeps the same face to its partner, much as the Moon keeps the same face towards the Earth. This means that each star spins on its axis at the same rate as the other, synchronized with the orbital period.

About 93 light-years away in the constellation of Perseus exists a famous interacting star system called Beta Persei, or Algol. The larger but dimmer of Algol's players is an orange subgiant. Its smaller partner is a blue star on the main sequence. Algol's two components swing around each other with unceasing regularity, once every 2.86 days.

Seen from Earth, Algol's orbital plane is almost edge on. So, twice every 2.86 days an alignment occurs. In the same way that the Moon sometimes eclipses the Sun, the larger star partially eclipses its smaller cousin. Outside of the eclipse, both stars are seen in their entirety and the system appears at its brightest. During the eclipse, the subgiant blocks some of the blue star's light behind it, and the system dims. About ten hours later, the subgiant moves away and restores the system to its original intensity. Once the stars complete half a turn, a less significant secondary eclipse occurs as the hotter star partially blocks the cooler one.

Algol is, therefore, an eclipsing binary. In fact, it was the first eclipsing binary star ever recorded, in 1669. (Cataclysmic binaries, X-ray binaries and RS CVn stars often exhibit eclipses, too.) This prototype is but one among hundreds of Algol binaries that show the same behaviour: a generally constant brightness for the majority of the orbital period, then a pronounced dip in intensity. This dip in intensity is attributed to an eclipse of the brighter and hotter primary star by the cooler and dimmer secondary.

Algols are semi-detached systems, in which the stellar separation is so tight that the gravitational pull of the primary on the secondary distorts the latter into an egg shape, as occurs in cataclysmic binaries or LMXBs. However, there is no accretion disc. The distorted orange star's atmosphere flows away from it and impacts the blue star directly on its equator. There, the impacting gas is flung off – a bit like water hitting a spinning ball – to form a messy and tenuous flow in orbit around the blue star.

"...the impacting gas is flung off... to form a messy and tenuous flow in orbit around the blue star."

From Earth, Algol binaries are always seen edge on to their orbital plane. However, in this depiction we see an Algol system from a slightly oblique angle and can witness the messy flow of turbulent gas that surrounds the hotter of the two stars.

*Scan the page to view animation*

An overcontact binary looks something like a gargantuan stellar dumb-bell, with the two stars enveloped in a common jacket of gas, rendering them the same colour.

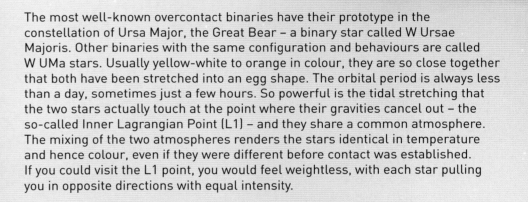

# CONTACT BINARIES
## STRETCHING OUT

So far, the kinds of interacting binary we have looked at share common traits. The secondary star is almost always distorted by the gravity of its compact companion (except for some HMXBs), and a bridge of glowing plasma, hundreds of thousands of kilometres long, flows from the secondary to the primary. Now we come to the systems where both stars are gravitationally deformed. These are often called contact binaries, but 'overcontact' is the better term.

The most well-known overcontact binaries have their prototype in the constellation of Ursa Major, the Great Bear – a binary star called W Ursae Majoris. Other binaries with the same configuration and behaviours are called W UMa stars. Usually yellow-white to orange in colour, they are so close together that both have been stretched into an egg shape. The orbital period is always less than a day, sometimes just a few hours. So powerful is the tidal stretching that the two stars actually touch at the point where their gravities cancel out – the so-called Inner Lagrangian Point (L1) – and they share a common atmosphere. The mixing of the two atmospheres renders the stars identical in temperature and hence colour, even if they were different before contact was established. If you could visit the L1 point, you would feel weightless, with each star pulling you in opposite directions with equal intensity.

As in the RS CVn binaries, the W UMa stars can have substantial spots – vast patches blanketing areas of their large surfaces. And like the Algol binaries, W UMa stars exhibit eclipses as each member periodically passes in front of its associate.

W UMa binaries are the cannibals of the stellar universe. In time, the larger of the two stars completely consumes its smaller, luckless collaborator, peeling away its atmosphere, layer by layer, until only the naked core remains. That, too, is gobbled up in a final unification, and a single star exists where previously there were two.

Proxima Centauri, as it might appear from the surface of a nearby world. If you were on Proxima now, Alpha Centauri A and B (top right) would be separated in the sky by only 0.16 degrees. But here Proxima is imagined at another epoch during its 500,000-year orbit, when the Alpha Centauri pair is much closer in by virtue of a highly elliptical orbit.

At last we leave binary systems to view additional stars. Like human multiple births, high-order stellar multiples are not common. Triple systems occur more frequently than quadruples, which in turn are more persistent than quintuple or sextuple stars, and so on. In a multiple system, the stars are usually bunched into pairs. So, in a triple there is often a close binary, with the third star orbiting much further out. Such a configuration is known as a hierarchical triple. In a quadruple star, two binary pairs orbit their common barycentre, and so on.

Examples of well-known multiple stars include Castor in the constellation Gemini, and Mizar in Canis Major. Each is a complicated throng of six stars. The nearest known multiple system is a triple, and it exists quite literally on our celestial doorstep. This is Alpha Centauri, or Rigil Kentaurus, one of the brightest stars in the sky.

Alpha Centauri, 4.36 light-years away, at first appears to be a binary system. The larger star is Alpha Centauri A, which is yellow and much like the Sun, but slightly bigger and brighter. Its cohort is Alpha Centauri B, which is smaller, dimmer and more orange. Owing to their elliptical orbit around each other, the distance between A and B varies from 11.2 AU to 35.6 AU (this binary would fit inside the orbit of Uranus). Further away is a third component. This is Proxima Centauri, the closest known star to the Sun and the red dwarf we met earlier (*see* pp40–41). Proxima orbits the Alpha Centauri pair at a vast distance of 13,000 AU – a meander of half a million years. This is so far-flung that it has been suggested that Proxima is not in orbit around the binary at all. However, the fact that all three stars are moving in the same general direction favours a weakly bound triple system.

**"Proxima orbits the Alpha Centauri pair at a vast distance of 13,000 AU – a meander of half a million years."**

# OPEN CLUSTERS
## STELLAR CROWDS

Lurking in the depths of space, clinging to the spiral arms of our Galaxy, are vast drifts of interstellar gas and dust, outweighing stars millions of times to one, light-years across. These are known as giant molecular clouds – the birthplaces of the stars. Usually, a molecular cloud splits up as it falls in on itself under gravity, each of the fragments forming a separate multiple star system or just a single star. But in some cases, the cloud fragments are so enormous that they give birth to not just multiple stars, but entire collections of them.

Our Galaxy contains an estimated 1,000 of these so-called open clusters, scattered around its spiral arms. They vary greatly in size and in their stellar cargoes. Some contain a few dozen stars in a region spanning a few light-years. Others are ten times larger and boast hundreds of members. In the densest, stellar objects are crammed in 10,000 times more tightly than in the neighbourhood of the Sun. If our modest yellow star were in the midst of such a cluster, its nearest neighbours would be less than a third of a light-year away instead of 4.3 light-years, and the night sky would be a much richer place.

A famous and very beautiful example of an open cluster is the Pleiades, or Seven Sisters (see pp104–105), in the northern hemisphere constellation of Taurus. The Pleiades cluster, easily visible to the naked eye even from a polluted city, is about 13 light-years across and houses perhaps 100 stars. We believe it formed about 80 million years ago. Open clusters do not last for ever. As a cluster moves around the Galaxy, encounters with neighbouring objects – such as other stars, other clusters and nebulae – stretch and distort the cluster, wrenching free some of its stars. While some clusters are billions of years old, in general they have much shorter lives.

> "Some contain a few dozen stars in a region of space spanning a few light-years. Others are ten times larger and boast hundreds of members."

A typical but fictitious open cluster of several dozen stars glitters in the depths of space like a clutch of celestial jewels. Two stars, perhaps a binary, are seen up close – one red and one blue – while the other members twinkle in the background.

# GLOBULAR CLUSTERS
## TIGHT BALLS OF STARS

Open clusters are big, but they have much larger cousins, known as globular clusters, which are roughly spherical in shape. They are enormous balls of shimmering stars, up to a million of them, crammed into a region 100 to 300 light-years across.

Unlike open clusters, globular clusters are truly ancient. They formed about 10 billion years ago, we believe, before even the Milky Way was complete. Globular clusters are therefore made almost entirely of geriatric stars. Because stars redden as they age, globular clusters have an orange tinge. And while open clusters stick to the spiral arms of our Galaxy, their globular counterparts lurk much further away. They surround galaxies in an enormous swarm called a halo, each of them on a highly elliptical path, greatly inclined to the plane of the Galaxy. Around 150 of them are known to orbit the Milky Way alone.

The closest and brightest globular clusters are visible to the naked eye. One called 47 Tucanae, in the southern constellation of Tucana, is 15,000 light-years away and about 150 light-years across – a glowing crowd of one million stellar objects. This cluster is famous not only for its relative brightness but also for its sheer density, one of the highest known. Deep in its heart the stars are stuffed in at a rate of about 1,000 per cubic light-year. Because of this enormous density, however, globular clusters are not safe places for life. Supernovae explosions shower nearby stars – and any associated worlds – with perilous levels of high-energy radiation. Collisions between stars are also somewhat common within globular clusters. Astronomers have found that some clusters contain anomalous blue stars, lost in a sea of old red ones. These objects – known as blue stragglers – are believed to be the result of stellar collisions, or they may be caused by the merging of two stars in a former close binary to form a single, hot star.

**"Deep in its heart the stars are stuffed in at a rate of about 1,000 per cubic light-year."**

On an imagined planet in the globular cluster 47 Tucanae, the sky would sparkle orange with the combined light of a million geriatric Sun-like stars. The average separation is just 0.1 light-years – so close that the Solar System would barely fit between them.

# 4 INTERSTELLAR FOG

Deep in the heart of the Milky Way, the galaxy to which our Solar System belongs, the region is ablaze with stars and glowing clouds of gas called nebulae.

# INTRODUCTION

For as long as we have had an interest in the skies we have known about nebulas – or, more commonly, nebulae. It's a word that comes from the Latin for 'cloud'. Initially, people used the term to describe any objects in space that looked a bit diffuse (fuzzy) either to the naked eye or with an optical aid. But as our understanding of physics and the Universe improved, and as the instruments at our disposal grew ever more complex, specialized and reliable, we came to realize that some of those blurry wisps were simply dense, previously unresolved clusters of stars, or even entire galaxies of stars beyond our own (*see* Chapter 5). True nebulae, as the term is now used, are different. They are vast patches of gas and dust that fill the swathes of space between the stars, in our Galaxy and in others. Collectively, they comprise what astronomers call the interstellar medium.

Nebulae are the Universe's places of outstanding natural beauty. But what are these enigmatic, multicoloured sculptures of light actually for? What do they do, how do they come about and how many different types are there? As we shall see in this chapter, nebulae are intimately linked to stars, star formation – and star death. Some are factories, bursting with the raw materials from which stars are created. Others are cosmic graveyards, the shattered remains of stars that once were – and will be again. They can be vividly luminous, energized by the light of nearby stars, or they may emit no visible radiation at all. And like the stars, nebulae exhibit staggering variations in size, mass and density.

# MOLECULAR CLOUDS
## STELLAR NURSERIES

Perhaps the most important components of the interstellar medium are molecular clouds. Dark, brooding and utterly frigid at temperatures of around -260°C, they are perfect breeding grounds for stars, and the planets that form beside them.

As their name suggests, molecular clouds are composed overwhelmingly of molecular hydrogen – a gas in which atoms of hydrogen are welded together in pairs to form a molecule. Molecular hydrogen, or H2, makes up around 88 per cent of a typical cloud's mass, with helium and trace gases making up another 11 per cent. The remaining mass, just 1 per cent, is made up of so-called dust. But do not imagine that this is the stuff that accumulates, unwanted, on your bookcases. Interstellar dust grains are far smaller, usually less than a thousandth of a millimetre across, and composed of silicates or rocky substances. With these dimensions they are comparable in size to the wavelength of light and, as such, are very good at blocking the passage of light. Thus molecular clouds are dark, emitting no visible radiation.

So how can astronomers spot them? Well, they show up as black patches when silhouetted against a brighter background. The Coalsack Nebula in the southern hemisphere is a classic example, as are the pillars of the famous Eagle Nebula shown here. And if, as is often the case, stars are embedded within these clouds, the dust grains absorb the light and warm up, emitting detectable infrared radiation.

Molecular clouds are very irregular, often assuming winding, serpentine shapes as a result of internal turbulence. The smallest ones are only 15 or so light-years across, with the mass of around a dozen stars. These are called globules. But most molecular clouds are giants, spanning hundreds of light-years and cramming in enough material to make a million stars like the Sun.

> "Dark, brooding and utterly frigid at temperatures of around -260°C, they are perfect breeding grounds for stars."

These classic molecular clouds, dubbed the Pillars of Creation, straddle several light-years and are part of the Eagle Nebula. This artwork shows the region closer to its actual colour than the false colour viewed in the famous Hubble Space Telescope image.

Protoplanetary discs like these are nurseries. Inside the dusty, pancake-shaped cocoons, hewn from initially much larger molecular clouds, violent processes are occurring that will one day create entirely new stars and planetary systems.

# 4.2 PROPLYDS
## PLANETARY NURSERIES

Just like clouds on Earth, their celestial counterparts the molecular clouds are irregular in shape and density. In some places, they are thin and wispy; in others, regions known as molecular cores, they can be much more substantial. And if the density exceeds a certain level, the molecular core begins to fall in on itself under gravity. It becomes smaller and denser, rotating faster and faster as it does so and flattening out. After about 100,000 years, in place of the molecular core is a gargantuan, opaque vortex of swirling gas and interstellar dust. This is a protoplanetary disc, or proplyd for short.

Proplyds are typically around 1,000 AU across. That's about 30 times the diameter of the orbit of Neptune in our Solar System. On their edges, they can be very cold, well below the freezing point of water. But in their central regions, where velocities, densities and friction are all far higher, temperatures can exceed 1,000°C. We saw in Chapter 2 how this leads to the formation of a protostar at the centre (*see* pp34–35), rocky planetesimals a little further out, and gaseous and icy planetesimals on the edges. This is known as the nebular theory of planetary formation. Indeed, the proplyd from which the Sun and its planets are believed to have been born is called the Solar Nebula. The theory neatly explains why stars end up in the middle and why the planets form further out. It also tells us how planets normally come to orbit in the same direction, spin in the same direction and orbit in the same plane.

Astronomers have discovered several proplyds in our Galaxy, the oldest being around 25 million years of age. In 1993, the Hubble Space Telescope zoomed in on a tiny portion of the Orion Nebula and found within it four stars still surrounded by protoplanetary discs. This has added considerable weight to the nebular theory of planetary formation.

**"Proplyds are typically around 1,000 AU across. That's about 30 times the diameter of the orbit of Neptune in our Solar System."**

"The gases within them are typically heated to temperatures approaching 12,000°C, and can weigh anything up to 20 times as much as the Earth."

These two lobe-like structures are regions of the interstellar medium that have been rammed by high-energy particles beamed along the axis of a protoplanetary disc, ionizing the gases and causing them to glow. We call this a Herbig-Haro object.

# 4.3 HH OBJECTS
## STELLAR SHOCKWAVES

Staying with protoplanetary discs for a while, we next encounter the odd-sounding Herbig–Haro objects, named after the two astronomers who first studied them in detail in the 1940s.

Via a mechanism that is still not fully understood, but probably involving magnetic fields, some protoplanetary discs emit powerful beams of subatomic particles, called polar jets. Hurtling outwards from the disc's centre at speeds of several hundred km/s, running parallel to the disc's spin axis, the jets collide violently with surrounding gases and create shockwaves that ionize them. Ionization is the process whereby a neutral atom – consisting of a positive nucleus surrounded by negative electrons – is broken apart. The fragments are called ions. Once ionized, the gas – now called a plasma – is in a high-energy state that is inherently unstable. This is because positive and negative ions, having opposite electric charges, tend to attract. So the ions in a plasma are constantly trying to recombine. When they manage it, light is emitted. The consequence is that, in response to the onslaught of ionizing shockwaves, the gases through which the particle beams travel will ceaselessly ionize and recombine, emitting beautiful displays of colour. We observe the phenomenon as a Herbig–Haro object, or HH object for short.

HH objects are a type of emission nebula, which we will meet soon. They resemble long, knotty, turbulent clouds terminating in lobes, and they can stretch across enormous distances – sometimes as much as a few light-years. The gases within them are typically heated to temperatures approaching 12,000°C, and HH objects can weigh anything up to 20 times as much as the Earth. Around 400 of them are currently known – but these nebulae are so ubiquitous in star-forming regions that their actual numbers in our Galaxy alone are likely to be thousands of times greater. The others are simply too far away to be seen.

# EMISSION NEBULAE
## GLOWING PLASMAS

"If the composition is suitably varied... an entire kaleidoscope of colours can manifest."

H H objects are an example of what astronomers call emission nebulae, because they shine by light that they themselves emit. Let's increase our field of view to look at general examples of this type of nebula.

Because molecular clouds are potential stellar factories, stars are commonly found within them. Usually these are young stars taking their first steps onto the main sequence and, as such, they are prodigious sources of ultraviolet light. This radiation can ionize interstellar gases effectively. As the ionized atoms recombine, they emit beautiful displays of colour that are the hallmark of emission nebulae. (A similar process generates neon light – a high voltage passed through the tube ionizes the neon gas, which emits light when its atoms recombine.) Gas clouds ionized by ultraviolet sources in sites of star formation are known as H II regions. The name (pronounced H-2) means ionized hydrogen (and is different from H2 – molecular hydrogen.) Other examples of emission nebulae include planetary nebulae and supernova remnants, both described later in this chapter.

Because of the universal predominance of hydrogen, which emits a characteristic red light called H-alpha when its atoms recombine, most emission nebulae are red. Other colours can be present, however, depending on the nebula's constitution and on the intensity of the ionizing source. If the composition is suitably varied, or if the ultraviolet radiation is powerful enough, an entire kaleidoscope of colours can manifest. That said, the colours that we see in photos of them are much more vivid than they would appear to our eyes – nebulae are too faint to trigger the cone receptors that colour our vision. Not only is photographic equipment more sensitive than our eyes to the colours in a nebula, but a camera's shutter can be left open for long periods, amplifying the light coming in and emphasizing the colours.

The Trifid Nebula, 9,000 light-years away in Sagittarius, is a staggering 50 light-years across. The red glow is H-alpha light, caused by the recombination of hydrogen, which is ionized by hot, young stars in the brightest part of the nebula. The dark regions are opaque molecular clouds in the foreground.

We stand on the surface of an imagined world orbiting one of the stars in the Pleiades star cluster. The sky is ablaze with the combined light of the cluster's other members. A nearby cold molecular cloud catches the light of these stars and manifests itself as a reflection nebula.

# 4.5 REFLECTION NEBULAE
## CELESTIAL MIRRORS

As we have learned, molecular clouds emanate no visible radiation of their own, while the warmer emission nebulae offer prodigious displays of dazzling colour as a result of intense ionization sources nearby or within them. Occasionally, however, the stars within a nebula may not be powerful enough to ionize, or the stars may reside outside the nebula, off to one side. In these cases, we might see what scientists call a reflection nebula.

As their name suggests, and unlike emission nebulae, reflection nebulae emit no light of their own. Instead, the interstellar dust particles within them catch the light of nearby stars and throw it back at us, like fog around a street lamp. This reflection process – known in astrophysics as scattering – is much more efficient at short wavelengths, towards the blue end of the spectrum, so blue is scattered more readily. Exactly the same process is responsible for our Earth's blue skies. In addition, most stars in and around nebulae tend to be young and, therefore, bluish. These factors conspire to rob reflection nebulae of the many colour variations that typify their cousins the emission nebulae, producing instead a mostly blue light. But they are still beautiful sights to behold.

> "...the interstellar dust particles within them catch the light of nearby stars and throw it back at us, like fog around a street lamp."

Around 500 reflection nebulae are known. Perhaps the most famous example is the nebula associated with the tight star cluster called the Seven Sisters, or the Pleiades. The Pleiades is a young cluster in the constellation of Taurus, less than 100 million years old. The stars are enshrouded in nebulosity that, at first, astronomers thought was part of the ancient molecular cloud from which the cluster formed. We now know, however, that this is an unrelated nebula in the interstellar medium which the stars are merely traversing on their ceaseless journey around the Milky Way Galaxy.

This example of a double-lobed planetary nebula looks like a celestial butterfly. The dying star's core, now a white dwarf, is at the centre. Often the expansion is spherically symmetrical, producing a simple shell of gas. Here, though, an unknown process – possibly caused by planets around the dying star or a binary partner – has shaped the nebula in a spectacular way.

# 4.6 PLANETARY NEBULAE
## DEATH SHROUDS

Hundreds of years ago, astronomers started to discover odd, fuzzy discs in the sky. Planets looked similar through early, primitive telescopes, so these objects became known as planetary nebulae. But they have nothing to do with planets. Now, with advanced observatories and with a better understanding of physics, we can appreciate these phenomena for what they truly are – the discarded envelopes of once mighty stars. They are a type of emission nebula.

A lightweight star such as the Sun, once it runs out of fuel, begins to expand as a result of changes in its core (*see* pp44–45). Without hydrogen, the nuclear furnace halts and the core contracts. It grows hotter and hotter until, at a certain temperature, the helium in it starts to undergo nuclear reactions. The core, alive again and dozens of times hotter than before, spews fresh energy outwards, and this pushes on the star's outer layers. The combustion of helium is highly dependent on temperature. So tiny changes in the core temperature equal vast changes in the reaction rate. This results in pulses rippling through the star, inflating the atmosphere even more until it detaches completely from the core. The core, now termed a white dwarf, ionizes these gases, and the result is a planetary nebula, one of the most beautiful lightshows in the Universe.

These nebulae can take on a bewildering variety of colours and shapes. Some are simply spherical shells, but more and more are being found with intricate structures such as lobes and jets. The envelope expands at a speed of around 20–30km/s and continues to do so for up to 50,000 years. After that time, with the nebula approaching a light-year or so in diameter, the gases are too far from the white dwarf for effective ionization and the nebula begins to fade. After billions of years, the ashes of the once-brilliant star are returned to the interstellar medium where, one day, they will form the raw materials for a new generation of stars.

> "The envelope expands at a speed of around 20–30km/s and continues to do so for up to 50,000 years."

# 4.7 SUPERNOVA REMNANTS
## SHATTERED STAR-STUFF

Stars that are more massive than the Sun also leave behind a nebula when then expire. But they don't go quietly. They go supernova! And in these cases, the energies and velocities involved are much more extreme than in the more sedate planetary nebulae.

When a star much more massive than the Sun exhausts the nuclear fuel in its core, there is a great, rapid implosion. Within a mere one second, the entire core of the star, initially larger than a planet, is compressed under gravity to the size of a city. It forms a tiny stellar corpse known as a neutron star (*see* pp50–51). Even more massive stars become black holes (*see* pp52–53). As a result of this implosion, shockwaves are sent outwards from the collapsing core through the surrounding stellar envelope, shredding it to pieces and throwing it into outer space at vast speeds, as much as 10 per cent of the velocity of light, or 30,000km/s. This is a supernova. So much friction is created within this expanding shell of gas that it easily reaches temperatures of millions of degrees Celsius. The result is a fractured cloud of shattered star-stuff, glowing with seething energy. This is a special kind of emission nebula called a supernova remnant.

Supernovae are vital. During the initial explosion, when temperatures run to extremes, nuclear reactions take place within the expanding gases that can occur under no other conditions in the known physical Universe. If it were not for supernovae, all the elements in the periodic table that are heavier than iron would simply not exist. There is no other way for them to form. All of the elements in your body that are heavier than iron – such as strontium, rubidium, zinc, bromine and lead – were forged in the furnace of an exploding star. The other reason why supernovae are so vital is that, like planetary nebulae, they enrich the interstellar medium. They return the gases borrowed when the star was alive to the cosmic gene pool, so that subsequent generations of stars – and in our case, humans – can flourish.

> "If it were not for supernovae, all the elements in the periodic table that are heavier than iron would simply not exist."

Where a star once shone, there now exists a tangled web of shattered star-stuff known as a supernova remnant, and the progenitor star's planets are left barren and sterile.

109

# 5 GALAXIES... AND BEYOND

If we could view the Milky Way from the outside, it might look something like this. The sprawling spiral is the home of our Sun – and every star we can see in the night sky.

# INTRODUCTION

So far we have examined planets, stars, clusters of stars and nebulae, but we have not yet seen how these objects fit together in the Universe's bigger picture. Perhaps unsurprisingly, there is incredible beauty, structure and order to the Universe. The members of the cosmic menagerie we have met are not randomly strewn throughout space. Instead, they belong to a gigantic collective – a super-sized construct called a galaxy. This galaxy, known as the Milky Way, is 100,000 light-years end-to-end and contains an estimated 200–400 billion stars, and quite possibly more planets even than that. It is flat like a disc and looks something like a spinning Catherine wheel. Astronomers call it a spiral galaxy.

The Milky Way is not alone. There are billions and billions of galaxies across the Universe. Each one is a sparkling island in the cosmos surrounded by the black sea of extragalactic space. Each has its own cargo of planets, stars, star clusters and nebulae. Most alien galaxies share the spiral form of the Milky Way. Others, called ellipticals, are gigantic, egg-shaped balls of stars. These are similar to globular clusters (*see* pp90–91) but a thousand times their diameter and a billion times their volume. Lenticular (or lens-shaped) galaxies are a mixture of these two classes. Then there are irregular galaxies, which have little, if any, coherent structure.

What lies beyond the galaxies? Well, even on these vast dimensions, gravity is hard at work. Its mighty influence bunches the galaxies into galaxy clusters, each containing dozens to thousands of individual members. On the largest of scales, even clusters of galaxies are subject to aggregation, being lumped into superclusters – clusters of clusters of galaxies.

You can learn more about galaxy clusters and superclusters later in this chapter. But let's start at the lower end of the cosmic scale and examine the Universe's building blocks – the galaxies – in detail.

# 5.1 SPIRALS
## WHIRLPOOLS IN SPACE

Whenever we think of galaxies, we likely picture the majestic spirals. They are among the most complex and photogenic galaxies, and they are the most common. Close to 60 per cent of all galaxies are estimated to belong to this class – or to a related one, the barred spirals, the group that includes our own Milky Way.

The spiral arms are rich in nebulae, so they are active sites of star formation and therefore somewhat bluish. There are stars between the arms, but they are cooler and dimmer, giving the 'space' a darker appearance. Seen from the side, a spiral galaxy has proportions somewhat like two fried eggs glued back to back. At the centre is its core, known as the nucleus or nuclear bulge, shaped like a slightly flattened sphere. It contains mainly old stars, reddened with age, and has little active star formation. Finally, on the outskirts of the galaxy, there is a spherical halo of old, red stars, many of which are confined to globular clusters (see pp90–91).

Like stars, planets and nebulae, galaxies vary greatly in scale. The largest known spiral galaxy, NGC 6872, spans as much as 522,000 light-years and crams in hundreds of billions of stars. The Milky Way, though by no means small, is moderate at around one-fifth of that size with an estimated cargo of at least 200 billion stars. At the other end of the scale are dwarf spirals – smaller versions of their full-sized cousins at one-tenth of the size, with about one billion stars.

As we have seen, astronomers classify objects. Spiral galaxies are given the designation SA followed by a letter a, b or c. SAa galaxies have tightly wound spiral arms and larger cores, while SAc galaxies have a looser spiral pattern and smaller cores. Astronomers used to think that these galaxies evolved from SAa through SAb to SAc, but we now know that galaxy evolution is more complicated than that.

"Close to 60 per cent of all galaxies are estimated to belong to this class – or to a related one, the barred spirals..."

This might be one of the most beautiful scenes imaginable in the Universe – the view of a spiral galaxy rising over the horizon on the surface of an alien world.

# BARRED SPIRALS
## THE DOMINANT SPIRALS

We learned earlier that the Milky Way is an example of a spiral galaxy. More precisely, our own Galaxy belongs to a subclass called the barred spirals. They make up more than half of all spirals, perhaps as many as two-thirds, so they are predominant.

Barred spirals are similar to their regular siblings in almost every way. Both classes have bright spiral arms encrusted with rich, star-forming regions full of young stars; both have nuclear bulges filled with geriatric, reddish stars; both are flat and rotate about the central hub; and both are surrounded by halos of globular clusters, each containing up to a million individual stars. The difference is that, in a barred spiral galaxy, the spiral arms do not begin at the centre and wind outwards. Instead, a thick bar of stars runs through the galaxy's core, and the spiral arms are attached to either end of this bar. The length of the bar is usually two to five times its width, and often it is marked with thick lanes of interstellar gas and dust where new stars are being born.

Again, as with regular spiral galaxies, astronomers have developed a classification scheme. Barred spirals with tightly wound arms and larger nuclear bulges are type SBa, while looser windings and smaller bulges earn them the designation SBb or SBc. The Milky Way, for example, is possibly classed SBb – but we cannot be certain, because we are unable to see the spiral structure from the outside. Galaxies that have weak bars intermediate between spirals (SA) and barred spirals (SB) and are called SAB. Because so many spiral galaxies are barred, some astronomers think that all spiral galaxies will host bars at some point in their existence, and that the bars are transient phenomena – possibly even cyclical. However, whether this is true or not is presently unclear.

*The Milky Way is a barred spiral galaxy of approximate class SBb, having moderately open spiral arms and a small central bar. This animation shows the Galaxy as it might appear from the outside. The yellow dot indicates the location of the Solar System.*

 *Scan the page to view animation*

# ELLIPTICALS
## GALACTIC FOSSILS

Spiral galaxies are spectacular and can be huge in terms of radius, but they are not the most massive galaxies since they are quite flat with limited volume. This is where elliptical galaxies dominate. They get their name from their appearance in the sky – they resemble fuzzy, oval blobs. What they lack in definition they more than make up for in terms of size and mass. The largest elliptical galaxies can stretch for millions of light-years and harbour in excess of a trillion stars. They are the true giants of extragalactic space.

Ellipticals have quite varied shapes. Some of them have a degree of symmetry, being shaped somewhat like a rugby ball or a squashed sphere. In these cases, they have cylindrical symmetry. Others are fairly spherical. In general, each of their three axes is of a different length, so their appearance will differ depending on the viewing angle. As with spirals, ellipticals are subject to a classification scheme. Those that look the most circular are of type E0; the more oval they are, the higher the classification number. Galaxies of type E7 are long and cigar-shaped. But the classification says little about the actual 3D shape of the galaxy. A galaxy of type E0, appearing circular to us in its orientation, might well be a long rugby ball when seen end on. There is no easy way to tell.

**"The largest elliptical galaxies can stretch for millions of light-years and harbour in excess of a trillion stars."**

Aside from size and shape, the other big difference between elliptical and spiral galaxies is that the former have little or no molecular clouds. Recall that molecular clouds are where stars are made. So, while spirals are alive with vibrant nebulae and newly forming stars, ellipticals are almost totally devoid of young stars and star formation. They are comprised almost entirely of ancient, reddened stars, billions of years old, and like spirals they are embedded in a swarm or halo of equally old stars that are gathered into globular clusters. In a sense, elliptical galaxies are fossils.

*This is how a typical elliptical galaxy might appear up close: a smooth orange oval, denser towards its core, with globular clusters and a few stars clinging to the outskirts. In the background are smaller, satellite galaxies, orbiting the main one.*

"Lenticular galaxies, being the mongrels of extragalactic space, share characteristics of both their elliptical and spiral brethren."

# LENTICULARS
## LENS-SHAPED GALAXIES

Having seen the most common types of galaxy – spirals and ellipticals – we come to a form that is a kind of hybrid of the other two. These are known as lenticular galaxies, classified S0. Their name comes from the fact that the original members of this class resembled optical lenses viewed from the side.

Lenticular galaxies, the mongrels of extragalactic space, share characteristics of both their elliptical and spiral brethren. They are disc-shaped, like spirals, and can have central bars, like barred spirals, but they appear to have no discernible spiral arms. They are more akin to ellipticals in that they contain very little interstellar gas, so they lack star-forming regions and are dominated by old, red stars, none younger than a billion years of age. They do, however, contain interstellar dust, which even ellipticals can lack. Some researchers interpret these facts to mean that lenticular galaxies are spiral galaxies which, having used up their cargoes of gas to make new stars, lost their spiral arms (where the star-forming nebulae reside). Support for this idea comes from the discovery of what astronomers whimsically call 'anaemic' spirals – true spiral galaxies that have faded arms, perhaps on their way to becoming lenticulars.

But there is another theory. Perhaps lenticular galaxies are the result of an extragalactic pile-up – two or more galaxies that collided long ago and subsequently merged into a new form. In fact, as we will learn overleaf, collisions between galaxies are pretty common – much more so than between stars. While the stars in a typical galaxy are separated like cherries placed hundreds of kilometres apart – unlikely to interact even with lifetimes of billions of years – the larger galaxies are scattered more like cherries less than one metre apart. It's a proximity that makes for a very dynamic cosmos.

These two galaxies are poised to merge. Already their close proximity has radically altered their shapes. Collisions trigger star formation, and successive mergers rapidly deplete the galaxies' interstellar mediums. This may be why ellipticals arising from galaxy mergers lack star-forming nebulae.

# INTERACTING GALAXIES
## COSMIC PILE-UPS

As we've already seen, compared to their overall sizes, galaxies reside very close to one another in extragalactic space. The reason for this is that galaxies are grouped together into close-knit bunches called galaxy clusters (*see* pp130–131). With this degree of crowding, interactions between galaxies are very common indeed.

Galaxies can interact with each other in various ways. In the simplest scenario, a small galaxy might pass close to a larger one – a spiral, say – and their mutual gravity will distort one of the spiral's arms. An alternative is that the smaller galaxy might dive right into a neighbouring one. Remember, stars are so widely separated that, in such events, very few actually collide. So, the smaller galaxy simply passes through the larger one like sand through a sieve, slows down a bit, and moves out the other side by virtue of its high speed. In an interaction such as this, the collision can send shockwaves through the target galaxy, compressing its interstellar gases and prompting vigorous star formation as these gas clouds collapse.

The most dramatic interactions between galaxies are mergers, where two galaxies meet head-on but neither has the momentum to keep going beyond the drama of impact. Instead, the galaxies buckle and warp around each other, then fall back towards each other to blend, eventually, into a single, larger galaxy. Astronomers call this 'galactic cannibalism', and some giant ellipticals are known to have grown to their great dimensions at the expense of smaller neighbours. We don't have to look too deeply into space to witness this. Several galaxies that orbit the Milky Way, notably the so-called Magellanic Clouds, are currently being 'cannibalized' and will one day merge with their larger parent.

Collisions and mergers are so common that astronomers think that they are the driving force behind galaxy evolution. Spirals form first, then collisions among them create messy-looking irregular galaxies (*see* pp124–125). Finally, with subsequent mergers, elliptical galaxies develop, growing ever larger as they suck in more and more of their smaller cousins.

" **...collisions and mergers are so common that astronomers think that they are the driving force behind galaxy evolution.** "

This is an impression of the Large Magellanic Cloud (LMC) – the largest of the Milky Way's satellites and usually classed as an irregular galaxy. Note the vibrant red nebulae where new stars are being born. The LMC is about 20,000 light-years across and lies roughly 160,000 light-years from Earth.

# 5.6 IRREGULAR GALAXIES
## EXTRAGALACTIC BLACK SHEEP

Galaxies that fit into neither of the other two major classes, ellipticals and spirals, are called 'irregular'. Despite this name, many of these extragalactic black sheep do have some sort of structure. Some of them have hints of spiral arms, or nuclear bulges that are way off centre. These are dubbed type-I irregulars, or Irr-I. Then there are those that seemingly have no organized shape or structure. They are chaotic scrambles of stars and nebulae, without symmetry, central bulges or spiral arms, and each one is unique. These are the type-II irregulars (Irr-II).

Irregular galaxies are fairly common, and important because they probably represent an intermediate stage in galaxy evolution, or perhaps a very early one. We saw on previous pages how gravitational interactions between galaxies, by virtue of their relative proximity to each other, are responsible for all sorts of distortions. Astronomers have created computer simulations showing how two initially separate galaxies evolve and then merge as they move closer together. The results of these simulations are strikingly similar in appearance to some of the Irr-II galaxies that we know of. So, some irregular galaxies can be considered 'missing links' in the chain of evolution from spiral to lenticular and/or elliptical. Other irregulars, however, might simply be young galaxies that have not yet fully developed their structure. Quite likely, both explanations are valid.

Our own Milky Way Galaxy plays host to a whole armada of smaller satellite galaxies. Most of these are dwarf ellipticals, but a few of them are irregular, no doubt because of gravitational interactions. Two of these satellites are the Large and Small Magellanic Clouds. You can see them clearly with the naked eye from the southern hemisphere, looking like two bits of the Milky Way that have broken off. Both show hints of a spiral structure – not to the untrained eye, though – and some astronomers prefer to classify them as barred spiral galaxies.

"... some irregular galaxies can be considered 'missing links' in the chain of evolution from spiral to lenticular and/or elliptical."

# RADIO GALAXIES
## COSMIC GUSHERS

We have now looked at each of the major types of galaxies. But what we have not yet remarked upon is the diverse range of electromagnetic radiation that galaxies can pump out. Among the most powerful are the so-called radio galaxies. These objects – almost always associated with elliptical galaxies – can generate as much as 100 trillion trillion trillion watts of power. That is more than one million times as powerful as the Milky Way, and much of that energy comes out in the radio region of the spectrum.

There is only one power source that can drive these powerhouses – and that is a supermassive black hole. We met black holes earlier (*see* pp52–53) – their gravity is so powerful that not even light can escape them. While the ordinary, stellar black holes cram the mass of 20 Suns into a sphere a few dozen kilometres across, the supermassive black holes that power radio galaxies are in another class entirely. They can weigh as much as one billion Suns and occupy as much space as an entire planetary system. As with stellar black holes, the supermassive ones are usually surrounded by accretion discs – gargantuan pancakes of debris made from shattered star-stuff. It is this in combination with powerful magnetic fields that makes supermassive black holes so conspicuous. Subatomic particles swirling around in the disc are accelerated to near light-speed by the magnetism and emit blistering amounts of what scientists call synchrotron radiation, at radio wavelengths.

The magnetic fields serve another purpose that makes radio galaxies so spectacular – that is, they can channel charged particles deep into extragalactic space, well beyond the host galaxy, forming long, narrow jets. These jets collide with the rarefied gas that exists between galaxies – called the extragalactic medium and comparable to the interstellar medium within galaxies – and inflate the gas to form huge, lobe-like structures that dwarf the emitting galaxy. These lobes are similar in appearance to the HH objects (*see* pp100–101), but on much, much bigger scales.

> **"There is only one power source which can drive these powerhouses – and that is a supermassive black hole."**

A radio galaxy, moving rapidly through the surrounding extragalactic gases, emits powerful jets of particles that plough into those gases and pump them up to form huge, radio-bright lobes. The lobes are swept back by friction with the extragalactic medium, creating structures like wakes on a pond.

# 5.8 QUASARS
## AT THE EDGE OF THE UNIVERSE

Radio galaxies are not the only ones in which monstrous black holes are thought to lurk. Indeed, our own Milky Way is known to host a supermassive black hole at its heart, and researchers have even detected radio lobes way above our Galaxy's rotation axis. Galaxies that are brighter than usual because of supermassive black hole activity are called active galaxies. Naturally, these include radio galaxies, but perhaps the most celebrated active galaxies are the mysterious quasars.

In the 1960s, astronomers began to find powerful radio sources that were well beyond our Milky Way. They had to be exceptionally bright, easily 100 times more powerful than a normal galaxy – otherwise, they would have been undetectable at such distances. Because they looked like stars they were dubbed QSSs (quasi-stellar radio sources – 'quasi-stellar' meaning 'star-like'). Nowadays, they are called quasars. The first quasar found, called 3C 273, is an astounding 2.5 billion light-years away; but others are at distances exceeding ten billion light-years. They are among the most distant objects in the visible Universe, yet their emissions have been found to emanate from a region just a few AU across. Most astronomers now agree that quasars are, like other active galaxies, powered by supermassive black holes. This explains why they can shine so conspicuously while also being so small and distant.

These days, astronomers suspect that most, if not all, galaxies play host to these voracious parasites. Why, then, are not all galaxies strong radio emitters, or as bright as quasars? The answer seems to be that, in many cases, the black holes are dormant. Their accretion discs have diminished so that they emit no particle beams or high-energy radiation. We have seen how interactions between galaxies cause gravitational perturbations. In very dense extragalactic regions, these perturbations can fling stars into the maws of supermassive black holes, creating fresh accretion and reigniting the high-energy emission. It is then that they flare up, and another active galaxy makes its presence known.

"The first quasar found, called 3C 273, is an astounding 2.5 billion light-years away; but others are at distances exceeding ten billion light-years."

Deep in the colourful heart of an active galaxy, we view the thick, dusty accretion disc (seen edge on) that powers a quasar. Powerful jets of charged particles shoot out of the disc's centre to head tens of thousands of light-years into extragalactic space.

*The Local Group galaxy cluster contains the Milky Way (top right), the Andromeda Galaxy (centre) and M33, the Pinwheel (bottom left). The animation shows the 3D distribution of at least 50 members within this cluster. The smaller galaxies have exaggerated scales; they are otherwise too small to see.*

*Scan the page to view animation*

# 5.9 GALAXY CLUSTERS
## LIVING IN GROUPS

Now that we've met the galaxy classes and the supermassive black holes that make some galaxies unusually active, it's time to zoom out and look at the environments in which galaxies live. They are called galaxy clusters – bunches of galaxies held together by gravity's all-encompassing glue.

On a clear night, a keen eye might spot the Andromeda Galaxy as a faint, oval smudge of light in the northern-hemisphere constellation from which it gets its name. Andromeda – or M31 to use its catalogue name – is a large barred spiral galaxy, bigger than the Milky Way, lying around 2.5 million light-years away. It is the furthest object in the Universe visible to the unaided eye. Meanwhile, a closer yet smaller and fainter spiral galaxy called M33, often referred to as the Pinwheel, lies in the Triangulum constellation. The Milky Way, Andromeda and the Pinwheel are the three biggest members of a galaxy cluster known as the Local Group. All the other participants, more than 50 in total, are dwarf galaxies mostly measuring no larger than a few thousand light-years across (compared to 100,000 light-years for the Milky Way). In total, the Local Group covers a region of extragalactic space measuring about 10 million light-years and weighing more than one trillion Suns.

That sounds impressive, yet the Local Group, with only three major galaxies, is puny compared to some galaxy clusters. One huge example, whose centre lies about 320 million light-years from the Milky Way, is the Coma cluster. It contains at least a thousand galaxies, including many spirals and giant ellipticals – the latter dominating its central regions. Countless other galaxy clusters exist at intermediate sizes.

The space between galaxies in a cluster is not empty. It is filled with an extragalactic medium called intracluster gas. Although very sparse, this gas exists at very high temperatures and is a powerful source of X-rays.

**"The Milky Way, Andromeda and the Pinwheel all belong to a cluster of galaxies known as the Local Group."**

# LARGE-SCALE STRUCTURE
## SUPERCLUSTERS AND VOIDS

There seems to be no end to the structure of the Universe and no limit to the amazing force of gravity. Its ghostly influence can link objects together on truly mind-boggling scales. Stars are grouped into galaxies, galaxies are combined into clusters and, incredibly, galaxy clusters are themselves subject to even larger aggregations. These are called superclusters – clusters of *clusters* of galaxies.

Just as the Milky Way belongs to the Local Group of galaxies, so the Local Group is but one of at least 100 galaxy clusters that comprise the so-called Local Supercluster. Also called the Virgo Supercluster, because many of its galaxies reside in the Virgo constellation, it stretches across a phenomenal 110 million light-years and houses tens of thousands of galaxies. It is somewhat flat, shaped like an ellipse. The Local Supercluster is just one of millions of superclusters in the known Universe.

Superclusters tend to chain together to create thread-like constructs, giving the Universe a somewhat frothy appearance. Between these delicate filaments are regions almost empty of galaxies, known appropriately as voids. They are roughly spherical and can stretch across 200 million light-years. That's 200 million light-years in which there is next to no visible matter at all and the Universe appears to be mostly empty.

This is, of course, an illusion. The Universe is brimming with planets, stars, nebulae, galaxies and galaxy clusters. Astronomers estimate that there are as many galaxies in the Universe as there are stars in the Milky Way – around 200 billion galaxies, totalling 40 billion trillion stars. Many stars seem to host several exoplanets, so the number of planets in the depths of the cosmos is likely to be higher still. To paraphrase Jodie Foster's character in the movie *Contact*, if we are the only life forms in the Universe, it would be an awful waste of space.

**"Astronomers estimate that there are as many galaxies in the Universe as there are stars in the Milky Way – around 200 billion galaxies, totalling 40 billion trillion stars."**

On the largest of scales, the Universe takes on a web-like appearance. Here, the white blobs are not stars; they are large galaxies or, more often, entire clusters of galaxies. Notice how they string together in chains. These are superclusters, and between them are the dark, nearly spherical voids.

**absolute zero**
Theoretically the coldest temperature, at which atomic motion stops (−273.16°C).

**accretion disc**
A swirling pancake of matter around some stars in binary systems and supermassive black holes in active galaxies. In binary stars, the matter is usually gas stripped from the larger star to encircle the smaller one. The disc around supermassive black holes is probably the remains of stars.

**accretion rate**
The speed at which matter passes through an accretion disc and accumulates onto the body at the centre of the disc.

**active galaxy**
A galaxy whose central black hole is presently active (consuming matter in its accretion disc). *See also* quasar; radio galaxy.

**asteroid**
An asteroid, or minor planet, is a chunk of the early Solar System, left over from its formation. It is made of rock, iron or carbon, or a mixture of these components.

**astronomical unit (AU)**
A unit used to measure interplanetary distances. One AU is defined as the distance from the Earth to the Sun, which equates to about 150 million km.

**atmosphere**
A shield of gas clinging to the surface of a planet or a star. The molecules that comprise a gas can move quickly. So if a planetary body does not have sufficient gravity to hold on to those gases, they will escape into space.

**atom**
An atom is a basic building block of matter. The nucleus at an atom's centre is made up of subatomic particles called protons and neutrons. Electrons orbit the nucleus.

**axis**
The imaginary line about which a body rotates. The Earth's rotation axis runs from the north pole to the south pole.

**barred spiral galaxy**
A spiral galaxy with a thick bar of stars that runs through its central regions. Spiral arms are attached to either end of the bar.

**barycentre**
The point between two celestial objects about which they would balance if joined with a large enough stick.

**binary system**
A system consisting of two separate components that are in orbit around their barycentre.

**black hole**
A compact object – usually a collapsed star – that is so massive for its size that its gravity is extremely powerful. Even light cannot escape its incredible pull.

**blue giant**
A very hot, massive, powerful star whose surface temperature (10,000°C–30,000°C) gives it a blue colour.

**brown dwarf**
A sort of 'failed star'. It forms like a star but its core is not hot enough to generate the nuclear reactions that would make it a star.

**cataclysmic binary/variable**
CV for short, this is a binary star system in which a white dwarf orbits a larger but less massive star, usually a red dwarf. The two objects are so close that gas flows from the red dwarf towards the white dwarf, usually (but not always) via an accretion disc. Examples of CVs include novae, dwarf novae and – where the white dwarfs are very magnetic – polars and intermediate polars.

**comet**
An icy body orbiting the Sun, left over from the processes that built the Solar System. As a comet approaches the Sun on its elliptical orbit, its surface ices begin to boil off and stream into space to form a spectacular tail.

**constellation**
A region in the night sky in which stars are grouped together to create a pattern.

**convective zone**
The region inside a star within which heat is transported to the surface via convection (exactly as boiling water in a pan rises from the base to the surface). Convective zones are where a star's magnetic field is generated.

**core**
The centremost and, usually, densest part of an object. Many objects in space have a core, including gas clouds, stars, planets and galaxies.

**dwarf planet**
A mini-world that orbits the Sun. Dwarf planets are massive enough for gravity to make them spherical, but unlike planets their orbits are shared with other, similar bodies.

**eclipse**
When a body in space passes directly in front of another, as seen from one vantage point, the alignment is called an eclipse.

**electromagnetic radiation**
A kind of energy emitted and absorbed by charged particles, travelling through space with wave-like properties. There are many kinds of this radiation, differing in terms of energy. The least energetic form is radio. Next are microwaves, infrared, optical radiation or 'light' (radiation we can see with our eyes), ultraviolet, X-rays and, finally, gamma rays. These are emitted by various space objects.

**electron**
A tiny subatomic particle, found on the outskirts of an atom, orbiting the nucleus. It has a negative charge that is exactly opposite to that of the positive proton. When atoms are ionized, some electrons are removed and become free-floating. They are then called negative ions.

**element**
Any material whose basic unit is an atom, such as hydrogen, helium and oxygen.

**elliptical galaxy**
An egg-shaped swarm of stars, bound by gravity. It can harbour up to a trillion stars and span hundreds of thousands of light-years. This type of galaxy lacks an interstellar medium (the gases needed to build new stars) so it contains mostly old stars and therefore has a reddish hue.

**equator**
The imaginary line that encircles a rotating body in space, perpendicular to and halfway down the rotation axis.

**exoplanet**
Any planet that orbits a star other than the Sun is called an extrasolar planet, or exoplanet. Scientists have found hundreds of them.

**extragalactic medium**
Extragalactic space, that which is found between galaxies, is not empty. It is filled with a hot medium called the intracluster gas or extragalactic medium, similar to the interstellar medium within galaxies.

**flare**
An energetic and sudden release of vast clouds of electrons, ions and atoms in a star's atmosphere, caused by magnetic activity. The most dramatic flares are those associated with flare stars.

**galaxy**
A vast collection of stars, gas, dust and planets. Galaxies come in many different sizes and types: spirals, ellipticals, irregular and lenticular.

**galaxy cluster**
The majority of galaxies belong to an even larger association, a group of galaxies known as a galaxy cluster. A cluster can house from 50 to more than a thousand individual members.

**gas giant**
A massive planet composed chiefly of hydrogen and, to a lesser extent, helium. It is a fluid world with no solid surface. In the Solar System, there are two gas giants – Jupiter and Saturn.

**giant**
The term 'giant' is used often in astronomy. A giant planet is one that is much bigger than the Earth (such as Neptune or Jupiter). A giant star is one in which hydrogen burning in the core has ceased, with fusion moving to a narrow shell on the core's outskirts. This puffs the star up to huge sizes, forming a red giant. *See also* subgiant; supergiant.

**globular cluster**
This is a cluster of very ancient, red stars found on the outskirts of many galaxies. It can span several hundred light-years and harbour up to one million stars.

**gravity**
Gravity is a force between two or more bodies, directly proportional to their masses and inversely proportional to the square of the distance between them. It operates on such huge scales that it dominates the Universe, holding together even the superclusters.

**helium**
The second simplest and second most abundant element in the Universe. Helium is created inside the cores of stars from the nuclear fusion of the simpler element hydrogen.

**Herbig-Haro object (HH object)**
This is an emission nebula associated with star formation. It results when newly forming stars beam subatomic particles into surrounding gases and ionize them. This causes the ionized gases to glow, creating a light-years-long, snake-like nebula along the path of the beam.

**hydrogen**
The simplest and most abundant chemical element, consisting of one proton (in the nucleus) orbited by one electron. Hydrogen is the chief component of nebulae, stars and giant planets. Inside stars, hydrogen is a fuel in chain reactions that generate an outward push and keep the star balanced against the pull of gravity.

**ice giant**
A massive, fluid planet rich in hydrogen, water, ammonia and methane. Uranus and Neptune are ice giants.

**infrared**
A form of electromagnetic radiation that we perceive as heat. It is emitted by many cooler objects in space, such as nebulae, young stars and planets.

**interstellar dust**
*See* interstellar medium.

**interstellar medium**
The space between the stars in the Milky Way is filled with clouds of material called the interstellar medium (ISM). Made up mostly of hydrogen, with some helium and also traces of rocky particles called dust, the ISM provides the raw materials from which new stars are made. When stars die, they replenish the ISM.

**intracluster gas**
*See* extragalactic medium.

**ion**
*See* ionization.

**ionization**
The process whereby atoms absorb so much energy that they break apart. When, for example, atoms in a nebula are bombarded with intense ultraviolet radiation from nearby stars, electrons are removed. They become free-floating and are referred to as negative ions, while the rest of the broken atom becomes a positive ion.

**irregular galaxy**
A galaxy that has little repeatable structure. Some irregular galaxies lack any kind of configuration at all and are little more than weird scrambles of stars and nebulae.

**Jovian mass**
'Jove' refers to the gas giant Jupiter, and a Jovian mass is the amount of material it contains (equal to 1.898 trillion trillion metric tonnes). It is used as a unit to measure the mass of exoplanets. So an exoplanet of 0.5 Jovian masses is half as massive as Jupiter.

**kinetic energy**
The energy that an object has by virtue of its motion. The more massive the object or the faster it moves, the higher the kinetic energy.

**lenticular galaxy**
A flat, disc-like galaxy. It is like a spiral but it has no discernible spiral structure. It has no interstellar gas.

**light-year**
A unit defined as the distance that light travels in a year, roughly equal to 9.5 trillion kilometres. It is used to measure vast expanses between stars.

**magnetic field**
Many space objects are magnetic. Immediately surrounding such an object and diminishing with distance is a region within which other magnetic or magnetizeable objects feel a force. This region is the magnetic field. Within the region, attracted objects are forced to move along the magnetic field lines towards the magnetic poles.

**main sequence**
The longest period in a star's life, during which nuclear reactions in its core convert hydrogen into helium. The power this generates keeps the star stable. Once the hydrogen runs out, the star is said to have left the main sequence.

**mass**
The quantity of matter in an object. Masses of stars or planets are often expressed in terms of other stars or planets. *See also* Jovian mass; solar mass.

**molecular cloud**
A large nebula, up to hundreds of light-years across, that is found in the arms and bars of spiral galaxies and may be common in irregular galaxies. It is made up of mostly hydrogen and helium, with traces of other elements and interstellar dust

particles. When a portion of a molecular cloud begins to contract under gravity, it leads to the formation of a star or a star cluster.

## molecule
A compound created by combining several atoms in a specific arrangement.

## moon
A moon is another name for a satellite – an object that orbits another, usually larger, body. Earth's only natural satellite is the Moon, while other planets have dozens of moons.

## nebula
A cloud of gas and, sometimes, dust usually confined to the spiral arms or central bars of a spiral galaxy. Nebulae are sometimes found in irregular galaxies.

## neutron
A subatomic particle that has no electrical charge. It is found in the nuclei of atoms.

## neutron star
When a very massive star reaches the end of its life, it collapses under its own weight, its core growing smaller and denser. Neutron stars are the densest objects in the known Universe.

## NGC
The NGC (New General Catalogue) is a list of astronomical sources. These sources are called NGC objects and are given these three initials followed by a number, such as 'NGC 7000'.

## nucleosynthesis
The process whereby atomic species are created within the core of a star via thermonuclear reactions. In a main-sequence star such as the Sun, nucleosynthesis creates helium from hydrogen. In massive stars, though, much more complex reactions can take place, creating chemical elements up to and including iron. Elements heavier than iron can only be synthesized in supernova explosions.

## Oort cloud
A hypothetical reservoir of comets surrounding the Sun at great distances, perhaps as far out as 50,000 AU.

## open cluster
A loose group of stars bound together by gravity. It can contain a few stars or hundreds, and can reach across several light-years.

## orbit
The path of a body moving under the influence of another's gravitational attraction. The time taken to complete one orbit is called the orbital period.

## planet
An object massive enough to be spherical and which encircles a star in an orbit clear of debris and other objects. It can be made of rock and metal, or chiefly composed of lightweight gases and ices. *See also* dwarf planet; gas giant; ice giant; terrestrial planet.

## planetary nebula
When a low-mass star such as the Sun exhausts the hydrogen in its core, it collapses under its own weight and jettisons its outer layers. The exposed core, now a white dwarf, ionizes the gases. They light up, forming a planetary nebula.

## planetesimal
An asteroid-sized fragment of rock, metal, ice or minerals created in a protoplanetary disc surrounding an infant star. It grows larger as it collides with and binds on to bits of debris, eventually earning the designation 'protoplanet'.

## plasma
A gas whose constituent atoms have been largely ionized. Stars and nebulae are made of plasma.

## proton
A proton is a subatomic particle with a positive electric charge. It is found in the nuclei of atoms.

## protoplanet
A young planetary body created during the formative stages of a planetary system inside a protoplanetary disc (or proplyd). It is large and round, like a fully fledged planet, but still growing as it continues to collide with neighbouring bits of debris. *See also* planetesimal.

## protoplanetary disc (proplyd)
A proplyd, which surrounds newly forming stars, is a flat disc of gas and dust within which planets are growing. The disc is created from molecular clouds as they collapse under gravity and spin up. *See also* planetesimal; protoplanet; protostar.

## protostar
A large, warm object on its way to becoming a star. It is found at the centre of a protoplanetary disc and can measure up to a few AU across.

## pulsar
A rapidly rotating neutron star that emits beams of radiation from its magnetic axis.

As the star spins, the beams flash across our line of sight and, like a lighthouse, the star appears to blink on and off.

## quasar
One of the most distant objects in the observable Universe, a quasar is an exceptionally luminous type of active galaxy. It is powered, like all active galaxies, by the accretion disc that encircles a supermassive black hole.

## radiative zone
The region inside a star immediately outside the core, within which energy is transported radially outwards by the flow of electromagnetic radiation particles, called photons.

## radio galaxy
An active galaxy that is very bright in the radio region of the electromagnetic spectrum. Driven by a supermassive black hole, this type of galaxy emits jets of fast-moving, charged particles from its central region along the spin axis. As these particles collide with the intracluster gas between galaxies, they inflate it to create vast lobes, which are also bright radio sources.

## radio lobe
*See* radio galaxy.

## red dwarf
The most common stars in the Milky Way (and probably the Universe) are red dwarfs. These low-mass main-sequence stars are smaller, less massive and cooler than the Sun and have a red hue. Many boast powerful magnetic fields and are great sources of flares.

## red giant
*See* giant.

## red supergiant
*See* supergiant.

## satellite
*See* moon.

## solar diameter
The diameter of the Sun, equivalent to 1.4 million km. It is often used as a unit to express the diameters of other stars. So a star measuring 0.5 solar diameters is half the diameter of the Sun. Similarly, a solar radius is exactly half of one solar diameter.

## solar mass
The quantity of material contained in the Sun – some 2,000 trillion trillion tonnes. This quantity is often used as a unit to express masses of other stars. So a star weighing 0.5 solar masses contains half as much material as the Sun.

## solar radius
*See* solar diameter.

## Solar System
The system that comprises the Sun and all that orbits it, including its planets, asteroids and comets. To the edge of the Oort cloud, the Solar System spans perhaps 100,000 AU in diameter. Planetary systems around other stars are often called solar systems (spelled in lower-case).

## spiral galaxy
The most common and beautiful type of galaxy, named because of its spiral structure. It is a flat disc of up to several hundred billion young and old stars.

The younger stars cling to and thus define the spiral arms, while the older ones inhabit the central regions and surround the galaxy in a vast halo of stars and globular clusters.

## stars
Gaseous bodies that shine by virtue of their own energy. They are created inside molecular clouds. Stars have nuclear reactions in their cores. These convert elements from one species to another (usually hydrogen into helium) and generate great energy in the process, which permits the star to shine. The Sun is a yellow main-sequence star.

## starspot
A cool region created on stars with surface temperatures comparable to that of the Sun or cooler. It often boasts an extensive convective zone inside of it, which creates a powerful magnetic field that can hinder the flow of gases on the surface of the star, lowering the local temperature and thus creating a starspot. The Sun's spots are called sunspots.

## stellar wind
A stellar wind (for the Sun, a solar wind) is an outflow of ionized or neutral gas emitted by the atmosphere of a star at speeds of between 10 and 2,000km/s.

## subatomic particle
*See* atom.

## subgiant
A star that has just left or is leaving the main sequence. The hydrogen in its core has almost run out, and thermonuclear reactions now take place in a thin shell surrounding the core, rather than within it. This generates more internal heat

than usual and puffs up the star's outer layers, resulting in a radius up to three times its size on the main sequence.

### superclusters
Clusters of clusters of galaxies, held together by gravity. They can hold tens of thousands of galaxies, and are the largest gravitationally bound systems in the Universe.

### supergiant
A huge, super-luminous star that has depleted the hydrogen in its core and evolved off the main sequence. The largest are the red supergiants, with some measuring 2,000 times the radius of the Sun.

### supernova
A huge explosion that marks the end of a very massive star. Briefly, a supernova can glow with more power than hundreds of billions of stars – outshining the galaxy in which it resides. The 'ashes' from the explosion mingle with the interstellar medium, where they will one day help to form new stars.

### synchrotron radiation
A special kind of radiation emitted by charged particles when moving under the influence of a magnetic field. In active galaxies, for example, this radiation is emitted by charged particles spiralling in the magnetic field that threads the accretion disc. In some cases the radiation emerges in the radio band of the electromagnetic spectrum, and we observe a radio galaxy.

### terrestrial planet
A world made up of solid substances: rock, iron and minerals. Mercury, Venus, Earth and Mars are the four terrestrial planets in the Solar System.

### thermonuclear reaction
A high-energy process in which atomic nuclei collide and merge, leading to the creation of successively larger nuclei. In a main-sequence star, hydrogen nuclei collide and bond to form helium nuclei.

### ultraviolet radiation
A very energetic form of radiation. It is emitted by many high-energy objects in space – including the Sun – but is mostly absorbed by the Earth's atmosphere.

### void
A vast, almost spherical region that exists between superclusters. It contains little if any visible matter and can span several hundred million light-years.

### white dwarf
The remains of the nuclear core of a main-sequence star. A white dwarf is created when a low-mass star runs out of hydrogen in its core, collapses under its weight and jettisons its outer layers into space to form a planetary nebula.

### X-ray
An extremely energetic form of radiation. X-rays are emitted by many high-energy space objects, such as the centres of accretion discs in compact binary stars.

### X-ray binary
A binary star system in which a compact object – a black hole or neutron star – orbits a large companion star. It is a very powerful source of X-rays. Systems in which the larger star is very massive are called high-mass X-ray binaries (HMXBs), while all others are called low-mass X-ray binaries (LMXBs).

# INDEX